Rehabilitation of the Human Bone-Muscle System

Edited by Adrian Olaru

Published in London, United Kingdom

Rehabilitation of the Human Bone-Muscle System
http://dx.doi.org/10.5772/intechopen.95131
Edited by Adrian Olaru

Contributors
Adrian Dumitru Olaru, Jinan Charafeddine, Samer Alfayad, Eric Dychus, Daoxiong Gong, Jianjun Yu, Dorin Copaci, Luis. Moreno, Dolores Blanco, Janeth Arias, Haobo Yuan, Lucia Beccai, Seonggun Joe, Federico Bernabei

First published in London, United Kingdom, 2022 by IntechOpen
IntechOpen is the global imprint of INTECHOPEN LIMITED, registered in England and Wales, registration number: 11086078, 5 Princes Gate Court, London, SW7 2QJ, United Kingdom

British Library Cataloguing-in-Publication Data
A catalogue record for this book is available from the British Library

Additional hard and PDF copies can be obtained from orders@intechopen.com

Rehabilitation of the Human Bone-Muscle System
Edited by Adrian Olaru
p. cm.
Print ISBN 978-1-80355-165-4
Online ISBN 978-1-80355-166-1
eBook (PDF) ISBN 978-1-80355-167-8

Meet the editor

Prof. Adrian Olaru graduated from the Faculty of Machines and Manufacturing Systems, *Politehnica University of Bucharest*, Romania, in 1974. From 1974 until 1990, he worked as a design engineer at the "Optica Romana" Enterprise, Bucharest, as well as an associate assistant in the *Department of Machine Building Technology, Politehnica University of Bucharest*. In 1990, Prof. Adrian became an appointed lecturer in the Faculty of Technological Systems Engineering and Management, Department of Machine Tools at the same university. Now, he is a full professor teaching the following courses: Dynamic behavior of Industrial Robots, LabVIEW Application in Modeling and Simulation of the Dynamic Behavior of Robots, Technological Transport Systems, Electrohydraulic Servosystems, Personal and Social Robots, Robots for Neurorehabilitation, and Vibration of the Virtual Prototypes of Industrial Robots.

Contents

Preface

There are significant differences between human muscles and artificial muscles, such as force rise time, contraction velocity, accelerations, moments that determine some variations in design, components, architectures, kinetics, and dynamics control. Artificial muscle plays an important role in exoskeleton robots, artificial upper and lower limbs, and remote control. A challenge to the research and engineering community is how to develop artificial muscles that mimic the performance of real human muscles.

The success of artificial muscles depends on both the inherent properties of ionic motion and the electrical voltage in concordance with the physical parameters. Future studies will need to address the feasibility of a prototype artificial muscle device in humans. Potential uses of artificial muscles in humans include breathing with an artificial diaphragm, movement of fingers, hands, feet, and facial resuscitation.

With respect to artificial muscle, the four basic prosthetic functions are elbow flexion, elbow extension, hand open, and hand close. The desirability of additional wrist and hand control is offered by the conductivity of ions during electro activation. Successful artificial muscles depend both on the inherent properties of ionic movement and voltage based on the physical parameters. Future studies will need to address the feasibility of a prototype artificial muscle device in humans. Potential uses for artificial muscles in humans include the establishment of breathing with an artificial diaphragm, movement of fingers and hands, and facial reanimation. Daily activities are a source of fatigue and stress for people with lower extremity spasticity. Thus, possible aids must be introduced while maintaining priority control by the patient. This work aims to develop such an application in the context of walking with an exoskeleton developed at the Systems Engineering Laboratory of Versailles (LISV). The application results are based on data recorded at the END-ICAP laboratory with gait sensors for healthy subjects, people with cerebral palsy CPs, and people who had a stroke. Our contribution is the proposal of a new method of neuromotor control for a rehabilitative exoskeleton. It consists in determining and assisting the motor instructions for the movements of a patient while retaining their expertise; the assistance as needed and the detection of its intention based on a fusion of information. The results show that the proposed index characterizes the relationship of the angle difference with a reference movement for each joint. It dynamically compensates for movements efficiently and safely. This index is applicable for gait pathology studies and robotic gait assistance. Other contributions present advances in the use of flexible Shape Memory Alloy (SMA)-based actuators for the development of upper limb rehabilitation exoskeletons that have been carried out by our research group. The actuator developed by our research group maintains the SMA wire characteristics (low weight, low cost, noiseless operation, compactness, and simplicity) and additionally presents flexibility and an increase in work frequency. These characteristics make its integration in rehabilitation exoskeletons provide the user with more comfort, ease of use, and freedom of movement. The chapter describes some different rigid and soft rehabilitation exoskeletons for different joints such as the elbow, wrist, and hand, in which this type of actuator has been successfully integrated. This gives the possibility to expand the research line with the actuated soft exosuits systems, in a future development perspective. Cervical laminoplasty is a posterior decompression surgery that was developed as an alternative to unsatisfactory results after laminectomy or corpectomy with multilevel fusion. Its usual indication is in cervical spondylotic myelopathy of more than three levels, associated with ossification of the anterior longitudinal ligament and multiple levels of spondylotic radiculopathy. Its contraindications are significant cervicalgia, kyphotic alignment, and segmental instability. Its advantages are that it maintains a high degree of mobility and avoids complications related to fusion. There are several techniques, the most important of which are Hirabayashi open-door laminoplasty and the Kurokawa double-door spinous process splitting laminoplasty. Complications after laminoplasty include the development of instability, gradual kyphotic deformity, restenosis, decreased range of motion, cervicalgia, and C5 root paresis, which is why the proper patient selection is imperative. Clinical results show that it is a safe and effective technique; there is no superiority of one procedure over another. The McKibben Pneumatic Artificial Muscles (PAMs) demonstrated in the book, which possess some compelling properties, are expected to play an important role in the advanced robots for endowing them with the ability of co-existing and cooperating with humans. However, the application of PAMs is still severely hindered by some critical issues. Focused on these issues, this chapter firstly presents a bionic design of a 7-DOF human arm-like manipulator in detail. This design takes the antagonized PAMs and Bowden cables to mimic the muscle-tendon-ligament structure of the human body, elaborately configures the DOFs of the arm, and flexibly deploys the routing of Bowden cables. As a result, the DOFs of the analog shoulder, elbow, and wrist of the robotic arm intersect at a point respectively and the motion of these DOFs are independent of each other for convenience of motion control and human-like dexterous manipulation. Another chapter presents a fSMC-ESO approach for human-like motion control of the PAM system. This approach views the model imprecision caused by the strong nonlinearity of the PAMs as a kind of internal disturbance and combines ESO with fSMC to observe and compensate for these disturbances. It is validated via experiments that this approach can realize human-like motion with expected robustness and tracking accuracy. The strong nonlinearity and the model imprecision problems are universally acknowledged as the main drawbacks of PAMs, and this approach is a feasible solution to it. Finally, some variants of PAMs that are aiming to amend or remedy the drawbacks of the PAM systems are discussed. Especially, a new approach that targets the dilemma of the output torque and the motion range of a PAM-actuated joint is proposed. Regarding the innovation of biomimetic cell culture scaffold, namely a 3D printed vibratory scaffold (3DPVS), has been proposed as a present-to-future novel product that is currently in the conceptual development stage. Design studies on concept generation of 3DPVS show high value, and one essential part inside this could dwell on establishing design methodological knowledge that has innovation merits. With its tools, TRIZ has proven value in creation and design innovativeness but has not yet been utilized for scaffold design at a mature level. In this chapter, we attempt to study and explore the design aspects of TRIZ and its most relevant tools in the context of 3DPVS, as well as preliminarily indicating a TRIZ-based methodology that could tailor the design aspects of 3DPVS. It also, to some extent, fills a gap in scaffold engineering and TRIZ literature and provides a comprehensive overview of a timely topic. In the past few years, vacuum-powered soft actuators have shown strong potential due to their promising mechanical performance. Indeed, they have been widely exploited in soft robots, grippers and manipulators, wearable devices, and locomotion robots. In contrast to inflatable fluidic actuators, the properties of the materials with which they are built have a stronger influence on the kinematic trajectory. For this reason, understanding both the geometry and morphology of the core structure, as well as the material characteristics, is crucial to achieving the desired kinetics and kinematics. In this work, an overview of vacuum-powered soft fluidic actuators is provided, classifying them according to morphological design, origami architecture, and structural instability. A variety of constitutive materials and design principles are described and discussed. Strategies for designing vacuum-powered actuators are outlined from a mechanical perspective. Then, the main materials and fabrication processes are described, and the most promising approaches are highlighted. Finally, the book discusses open challenges in enabling highly deformable and strong soft vacuum-powered actuation.

Prof.univ.Ph.D.Eng. Adrian Olaru
Department Robotics and Production Systems,
University Politehnica of Bucharest,
Bucharest, Romania

Section 1

Introduction

Chapter 1

Introductory Chapter: Introduction to Rehabilitation

Adrian Olaru

1. Introduction

Rehabilitation robotics is part of an approximately new and constantly growing field, especially in the clinical environment. Pioneering technologies have been discovered since the late 1980s and early 1990s, with the recovery induced by sensory-motor function training applied to animals with central nervous system (CNS) damage [1].

The use of machines in rehabilitation, however, comes much earlier, through a patent proposed in 1910 by Theodor Büdingen, which develops an electrically operated machine in order to guide and support walking movements for patients with heart problems.

In 1930, Richard Scherb developed the first man-operated, cable-operated mechanotherapy machine to mobilize joints, followed by the development of the first robotic rehabilitation system based on continuous passive movement, which was dependent on the patient's contribution due to its rigid connection system.

The applicability of robots in the therapeutic field has been introduced since the 1970s with the advent of the first monitored exoskeletons and equipped with pneumatic, hydraulic, or electromagnetic batteries for position control. These included advanced features such as ankle flexion/extension, adduction/abduction of the hip required for increased stability, and exoskeleton movement that could be carefully controlled by the therapist through the movement of one's own body (similarly connected exoskeleton) [2]. Thus, the first system invented for robot-assisted therapy for post-stroke patients was based on a rigid industrial maneuver, which did not allow interaction with patients, but only moved a tampon in different areas for patients who had to touch it.

In 1989, with the advent of MIT-MANUS equipment, first tested in 1994, a new era of neurorehabilitation robotics took place. This plantar manipulator had a low impedance (resistance detected during movement between the human-user interface and the robotic system) so that it provided increased stability of the upper limb by unloading it against the weight, being excellent for the severity of deficits [3, 4].

After a few years, the new generation of force-controlled devices used in bimanual gripping and lifting movement is also introduced, which led to the initiation of a more advanced control interaction, for the passive movements of patients with severe disabilities and for the assisted active movements of patients with moderate disabilities.

At the same time, mirror image motion enabler (MIME) appears on the market, which is equipped with a rigid industrial robot and offers support in performing paralyzed limb movements, but through the movement diffuser, it controls the healthy limb (mirror image therapy) [5].

For the extremity of the lower limb, rehabilitation robots have been developed since 1994, when the Locomat neurological recovery equipment was launched, which supported the body weight with the help of a robotic walking orthosis, simultaneously with its training on the treadmill, as well as the appearance of the GIT Trainer equipment with a similar concept, but with a final effect design [6].

Lokomat is an exoskeleton-type system, essential in the process of relearning independent walking, which involves activating the balance by lateral movements and rotation of the pelvis, with the help of a support platform of the center of gravity.

During therapy, attractive exercises are performed, which stimulate the patient's effort and motivation by including competitive elements and a data storage system. All aspects of gait recovery are monitored, such as the path of the foot during the gait cycle (initial contact and detachment) and the length of the step, so that the option to increase the level of movement is beneficial in the relearning process [7, 8].

The following years after these discoveries, a wide range of rehabilitation robots for the lower and upper limbs were brought, being classified globally according to their complexity as follows: stationary exoskeletons, on the ground; effector skeletons with remote operation through a device called Gripper; portable exoskeletons.

The first two categories of exoskeletons are well defined, compared with the portable exoskeletons that are currently undergoing clinical testing.

Therefore, rehabilitation robots must allow a physiological stimulation of the limbs in performing training with the functional movements of the affected segments, but also on the stimulation of peripheral receptors for functional training in performing the step.

With this historical and clinical background of real importance in the field of rehabilitation robotics, the neurophysiological basis is underlined in the design of recovery equipment that will follow in the future developments [9, 10].

Currently, there is interesting equipment on the market in the field of rehabilitation robotics, ranging from hard, fixed structures to light structures that are customized and directly operated. The combination of robotics with non-invasive and invasive brain-machine interfaces or neuro-prostheses in order to determine independence in everyday life is also at an early stage [11, 12].

Bionic technologies make possible the human connection with the interface of computer systems. They are called brain-computer interfaces (ICCs) also known as BCIs. These interfaces are either input, when they generate signals related to the nervous system or output, when the signal triggered by the wearer's nervous system was recorded and processed by the ICC, in order to control a computer or a robotic system.

The development of the first exit ICCs is aimed at recovering motor and communication skills in patients with spinal tract disorders and muscular dystrophies with disabilities, but also having amyotrophic lateral sclerosis. The interfaces connect to robotic manipulators and robotic wheelchairs, and also to programs that allow the recording of clamping actions and movement all based on the analysis of thinking activities.

Fundamental in the design of the output ICC, the first systems used the type of nerve signal based mainly on electrical signal. The electrodes used to detect electrical signals can be placed on the scalp in the case of Electro Encephalo Graphy—EEG, on the area of the cerebral cortex or deep in the brain tissue. In this way, the concept of an ICC is the basis for detecting electrical peaks in the brain, digitizing them and translating those actions that the brain does, thus providing control over the devices around us, with just one thought [13–15].

The three main categories of ICC are invasive—those implanted in the brain and provide a good reception in the ability to gather essential information about neurons and their activity; partially invasive (Neuralink)—placed in the skull, but not directly on the brain, which are safer, more effective, and less intense surgically; non-invasive—in the form of rubber hats and helmets with electrodes. They are easy to use, but because they are located away from the brain, electrical signals are not as accurate as brain implants.

Since 1969, scientists have successfully completed the project in which monkeys moved a needle on the dial of a computer, using only brain signals. In 2008, the brain was used to control the arms of a food-producing robot. The monkeys move the robot's arms after receiving feedback to the area responsible for controlling the arms, like a real arm.

In 2006, a study was conducted that allowed a robotic limb to be manipulated using the brain computer interface (BCI) with electrodes inserted into the user's motor cortex, a program that tracked human interaction with household appliances [16].

Following the fact that in 2014 people successfully used brain signals to control robotic arms and legs, the proof is shown by Juliano Pinto, the paraplegic who used a mind-controlled exoskeleton at the World Cup in Brazil.

Neuralink is part of a partially invasive ICC project and uses robotic tools to implant tiny wires into the brain. At present, it is still necessary to make small holes in the scalp, but in the future, they will be made with a laser.

The wires implanted in the form of chips are connected to an external device, called the link, which translates the signals provided by the brain and sends them to other devices. Currently, studies are being done on animals and will be performed on people, especially those with medical needs, but also for the use of other devices such as smartphones.

Cyberkinetics developed in 2004, the first modern BCI, a brain implant called Brain Gate, through which carriers were able to connect their brains to the computer, and in 2012, the company demonstrated the effectiveness of the device by controlling a robot arm by the brain. People with spinal nerve injuries enjoy limb control by using these brain signals [17].

Thus, the development of non-invasive ICC applications on EEG targets a wide range of potential users. The current projects of companies interested in the study of BCI technologies are limited to non-invasive devices, such as: emotional electroencephalograph headphones that take the signals from the brain, analyze them, and provide the consumer with information about how to use the devices; Neurable—a non-invasive BCI, hands-free, voiceless headset with VR technology to train the mind in virtual limb adaptation.

In the area of prostheses and neuro-prostheses, the research was performed on invasive solutions of brain computer interfaces, thus aiming to intercept muscle activity by means of electrodes placed on the user's skin (electromyography).

Such prosthesis is represented by the robotic hand Michelangelo, produced by the German company Otto Bock HealthCare, through which the subjects controlled the movements of the robot limb with the help of the activity of their own muscles. This is the first electronically operated device that mimics the natural movements of the human hand and is used for a variety of everyday tasks (cooking, ironing, brushing, driving). The material of the prosthesis is anthropomorphic elastic adaptable for each user, and the built-in electrodes detect the movements of the healthy muscles of the wearer, which are interpreted at the software level by electromyography.

Since 2008, Advanced Arm Dynamics has used prostheses on both civilians and military amputators in the United States and the United Kingdom, and will be deeply involved in their development and testing in the coming years [18]. This research is still in full swing, and technology is evolving from functional recovery devices to care devices, assisting to make up for existing sensory-motor impairment.

Subsequent advanced approaches for actuation, detection, and control make simple devices robust with clinical and home applicability.

In the future, wearable devices require the ability to adapt, but also to reduce the support of recovery plates, thus compensating for chronic deficiencies.

Developed robotic control technologies to assist people with disabilities have provided additional support by manipulating objects with the help of robotic arms, installed on desks or benches for the purpose of handing over objects.

Physical interaction systems with other devices, such as PAM-Aid (Personal Adaptive Mobility) and NavChair (developed by the University of Michigan), have also been developed since 2001.

NavChair has a wheelchair, equipped with a mechanism to control movement, and avoids obstacles, determined by the activity of detection and control of position sensors. Such a support can be provided while driving and robots such as PAM-Aid, which have the ability to avoid obstacles, but also to brake independently in the event of unforeseen obstacles.

Therefore, future approaches offer a number of challenges for both therapy and robotic care, many requiring a deep understanding of the neurophysiological systems underlying sensorimotor functions, as well as collaboration with industry and non-governmental organizations.

Also, in the future, the design of exoskeletons will involve much more work, with better materials and with easy design and control, and with the improvement of nanotechnology, we will move toward an era in which cyborgs will become a necessary reality [2].

Thus, the study and future research are related to the biomechanics of the spine, the morphology of the locomotor system, gait; however, neuronal control remains extremely important.

2. Anatomy of the spine and nervous system, diseases, and methods of recovery

The study of human anatomy has had a long history, dating back to antiquity, through the embalming of corpses, which led to the first descriptions of the brain and its membranes and later in the Renaissance, known as the period of anatomy today.

During the two great periods, the study of the human skeleton led to the discovery of the 200–220 bones, of which it is composed of the following main structures: spine constructed of 33–34 vertebrae (7 cervical, 12 thoracic, 5 lumbar, 5 sacral, 4 coccygeal); chest with 12 pairs of ribs; skull (29 bones); upper limb bones (64 bones); bones of the lower limbs (62 bones).

These structures that make up the human skeleton are directed, from a functional point of view, through a complex system of the organism called the nervous system, which offers the possibility for the whole organism to communicate with the environment.

In turn, any damage to the existing structures of the human skeleton causes the development of diseases of the spine with the need to apply locomotor or neuromotor rehabilitation treatments.

In developing countries, spinal cord injuries are one of the leading causes of death in the first 40 years of life and are also responsible for increasing the number of disabilities due to associated trauma (pelvic fractures 18%, long bone fractures 14%, craniocerebral fractures, thoraco-abdominal injuries) [18].

The main cause of acquired disability worldwide remains the field of neurological disorders (strokes, craniocerebral trauma, and neurodegenerative diseases), which requires methods and techniques of recovery in the field of neurorehabilitation.

Following the synthesis provided by Reinberg, the spine is an extremely important segment from a functional point of view, consisting of 33–34 bones, 344 joint surfaces, 24 intervertebral discs, 365 ligaments, and 730 dotted areas of insertions and origins of muscle bundles. The spine is divided into four regions consisting of a fixed number of vertebrae: the cervical region (neck) consisting of seven cervical vertebrae with a role in ensuring the mobility of the head. This function is provided by the morphological peculiarity of the cervical vertebrae with a larger transverse diameter of the vertebral bodies. The thoracic region (thorax) consists of 12 thoracic vertebrae. The position and oscillation of the center of gravity is represented by the particularity of the vertebrae in this area, through which diameter is larger in the anterior-posterior area. The lumbar region, with five vertebrae and large diameter in the transverse area of the vertebral bodies, offers increased mobility. The sacral region (pelvis) with five sacrococcygeal vertebrae and welded is called false vertebrae. The coccygeal region is represented by an axillary-like triangle, with the tip down and formed by the union of 4–5 vertebrae.

The vertebral column shows the articulation in the upper floor of the cervical vertebrae with the skull, and in the lower floor, it articulates through the sacrum with the coxal bones [19].

The spine occupies an important place in human physiology, being the central axis of the human body that fulfills the mechanical aspects: rigidity and elasticity.

Overall, it consists of three normal curves presented as follows: cervical curvature with anterior convexity; dorsal curvature with posterior convexity; lumbar curvature with anterior convexity; sacral curvature with posterior convexity.

The presence of these physiological curves gives the spine an increased resistance of up to 10 times higher for heavy loads.

The movements of the spine are varied and complex as a result of the cumulative movement of all the connecting elements: intervertebral diarthrosis, intervertebral discs, intervertebral symphysis, including the movements of the intervertebral joints (flexion, extension, lateral tilt rotation).

From the point of view of the articulation of the vertebrae, this is done through a symphysis with the interposition of the fibrocartilaginous disc, called the intervertebral disc, which in turn consists of a nucleus pulposus with a content of 90% water and fibrous ring.

In the back area, the upper and lower vertebrae articulate with each other through simple arthrodesis thus making the sliding motion possible.

The main joints that make possible the varied movement of the spine are as follows: (i) The costovertebral joints with an important role in respiratory biomechanics ensure the union between the vertebral extremities of the ribs with the spine; (ii) the joints of the rib head are flat diarthrosis that connect the joint face of the rib head and the body of two adjacent vertebrae; (iii) the costotransversal joints are diarthrosis that connect the articular face of the costal tubercles for the first 10 ribs and the costal face on the transverse apophyses of the first 10 thoracic vertebrae; (iv) the sternocostal joints formed at the union of the cartilages of the 2–7 ribs with the costal incisions of the sternum.

The stability of the spine is provided by the presence of the posterior longitudinal ligament placed posteriorly by the vertebral bodies, the anterior longitudinal ligament placed anteriorly by the vertebral bodies, as well as the presence of the spinal muscles, located in two superficial and deep layers. The most protected area from a muscular point of view is the portion of the lumbar area.

Due to these multisegmental structures, the spine performs a complexity of static and dynamic functions, such as: (i) static function—resistance in the application of the force of gravity and in the support of the head in position; (ii) dynamic function—complexity in performing active and passive movements; (iii) neuroprotective function—overlapping vertebrae and forming the spinal canal in which the spinal cord is located.

From the point of view of the function performed, the nervous system is present, as one of the most complex functional systems of the human body, without which it is not allowed to communicate with the external environment.

The anatomical division of the nervous system involves its division into several aspects: (i) the central nervous system consisting of the spinal cord located in the spinal canal and in the brain in the cranial box. The encephalus consists of the brainstem (bulb, bridge, midbrain) cerebellum, diencephalon (thalamus, metatalamus, epitalamus, hypothalamus, and subthalamus) and telencephalon, and (ii) peripheral nervous system represented by spinal nerves and cranial nerves.

From a functional point of view, the nervous system is divided as follows: (i) somatic nervous system, with direct communication with the external environment; and (ii) vegetative nervous system, with direct communication with the internal environment and in turn divided into central nervous system and peripheral nervous system [20].

From a structural point of view, the nervous system is structured from (a) gray matter that in turn consists of the bodies and dendrites of neurons, placed like a nucleus; (b) white substance consisting of myelinated axons arranged in cords, tracts, and bundles; (c) reticular formation and ependyma.

The nervous system consists of neurons and neuroglia with the function of supporting neurons.

The spinal cord is located in the neural canal of the spine and is an elongated portion of the central nervous system, occupying 2/3 of the upper portion of the spinal canal. Its length is 42–45 cm and extends from the upper edge of the atlas to the level of the vertebrae L1–L2, having in the upper part the spinal bulb and the lower medullary cone. The spinal cord is covered by the spinal dura mater, the spinal arachnoid, and the spinal pia mater and separated by the subdural and subarachnoid space, respectively.

Depending on the exit points of the spinal nerves, the spinal cord is divided into several parts as follows: cervical portion with cervical areas 1–8; thoracic portion with thoracic segments 1–12; lumbar portion with lumbar segments 1–5; sacral portion with sacral segments 1–5; coccygeal potion with coccygeal segments 1–3.

The spinal nerves originating in the spinal cord represent the communication of nerve flow from the external environment to the spinal cord. These pathways are formed by joining the nerve fibers of the nerve roots in the anterior and posterior area after passing through the three sheaths of the spinal cord [10].

The spinal nerves are mixed, made up of sensory, motor, and associative fibers.

The presence of the 31 pairs of symmetrical and meta-symmetrical distributed spinal nerves is as follows: 8 pairs per cervical area; 12 pairs on the thoracic area; 5 on the lumbar area; 1 on the coccygeal area.

Anastomosis of the branches of the spinal nerves causes the formation of distributed nerve plexuses and these depend on the irritated area: (a) The cervical plexus formed by welding the anterior branches of the first four cervical spinal nerves; (b) gill plexus formed by welding the anterior branches of the last four pairs of cervical spinal nerves (C5–8); (c) the intercostal nerves with 12 thoracic nerves, which form muscle and skin branches; (d) lumbar plexus formed by welding the anterior branches of the T12 and the first four lumbar nerves. It innervates the area of the abdominal wall, obturator muscles, thigh, genitals, leg, and leg; (e) duplex plexus formed by joining the anterior branches of the fourth sacral nerve and innervates the pelvic muscular and visceral area; (f) the coccygeal plexus formed by the union of the anterior branches of the fifth sacral nerve and the coccygeal nerve with the innervation of the ischia coccygeal area.

3. Displacement and range of motion achieved in the spine and musculoskeletal system (biomechanics)

The movement and range of motion achieved in the spine include all the movements of the kinematic elements of the human body through a reference system.

This reference system is the fixed or mobile benchmark against which all the positions of a system under evaluation are measured. Thus, a motion reported and compared with a fixed reference system is called an absolute motion, while the measured motion at a mobile reference system is called a relative motion.

The human body can perform simple transitional movements and rotational movements, all due to the complex and simple joints of the spine, but also of the musculoskeletal system. The rest of the body movements, the pivoting movements and the roto translational movement are achieved by completing the simple movements both in space and in plan.

In the biomechanics of the spine, the origin of the reference system is usually in the center of gravity of the human body, which moves with the human body and is called the relative or cardinal reference system.

The biomechanics of the spine have certain degrees of freedom offered by different segments: vertebrae: C3-C4-C5-C6-C7 perform the flexion movements—extension of the front, back and side of the upper limbs, being the most flexible vertebrae; the T1-T10 vertebrae are the most fixed by their articulation with the ribs and later they articulate with the sternum; the T11-T12 vertebrae are mobile because they have a joint with false, mobile ribs; the L1-l5 vertebrae are also mobile. They allow flexion extension of the lumbar area.

The sacral and coccygeal vertebrae are fixed, immobile, and welded.

In the case of spinal injuries, the emphasis is on their physiological and anatomical features, as follows: (i) The cervical vertebrae are smaller and thinner than the lumbar ones, these being the most affected in vertebral traumas; (ii) the transitional position between the mobile and the immobile area is extremely important, because the cervical vertebrae are at a point that has a kinematic energy applied in the collision as the highest; (iii) injury of the spinal cord in the cervical area will cause tetraplegia, and injury in the lumbar area paraplegia.

The movement of the joints of the locomotor system involves knowledge of the anatomical systems about his components and how human locomotion is performed.

The locomotor system has a set of anatomical systems, including the osteo-articular system, the nervous system, and the muscular system.

4. Diseases and recovery of the spine

The most common spinal disorders are as follows: (i) Herniated disc is an effect of a neurological nature characterized by the sliding of the nucleus pulposus of the intervertebral disc along the spinal cord with compression of the spinal nerves and the determination of painful radiculopathy. It can be as the result of a spinal cord injury, common in the lumbar and cervical region; (ii) discopathy is a condition of the intervertebral disc with aging and deterioration of the disc with reduced flexibility and mobility; (iii) lumbosciatica is a condition manifested by pain, tingling, numbness caused by irritation of the sciatic nerve as a result of narrowing of the spinal canal (spinal stenosis), osteophytes, and arthritis of the vertebral joints; (iv) Lumbago is a contracture of the paravertebral and lumbar muscles as a consequence of irritation of the lumbar nerve by displacement of the disc nucleus. It is installed suddenly and painfully; (v) spinal stenosis occurs as a result of degenerative diseases that cause changes in the joint areas, which decreases the space of the spinal canal. The damage occurs at both the cervical and the lumbar levels.

The recovery of these diseases of the spine is done through recovery therapies such as massage, ultrasound therapy, electrotherapy, laser, teak therapy, shockwave, cryotherapy. For the use of these therapies, medical recovery equipment is used to reduce the inflammation, pain, and edema present in the joints, muscles, and ligaments of the spine.

Over time, it has been found that the main diseases of the spine that cause long-term physical disabilities in adults are stroke and spinal cord injury.

Regarding the survival rate, approximately 10 million people survive trauma after stroke and over 250.000 survive after spinal cord injuries.

This is due to post-stroke activation deficiencies, which limit their self-reliance. Also, as a result of injuries to the central nervous system, there are side effects such as spastic muscle tone.

Compensating for sensorimotor deficiencies is spasticity, an aid in restoring lost function. Thus, spasticity has been clinically proven by studies that can be used to partially reduce the loss of limb activation in mobile patients, and spastic legs may act as a support for the body while walking or in maintaining a stick-like position [7]. These statements are valid only for moderately affected patients, while for those with severe disabilities, spasticity and muscle cramps require pharmaceutical interventions due to their exaggerated intensity.

Statistically, at European level, there have been 200 cases of stroke in 100,000 people reported. Of these records, 10% integrate into society, 40% remain with moderate sequelae, 40% have severe sequelae, and 10% are unrecoverable.

Recovery from conditions such as stroke and multiple sclerosis and last but not the least spinal cord injuries have a difficult goal to define, because conventional therapies hardly separate the spontaneous recovery of function that occurs with rehabilitation treatment. In post-stroke recovery, the motor recovery process takes place mostly in the first 3 months and at 6 months with little measurable progress.

Depending on the time of the stroke and the time of recovery, several phases can be defined: (a) initial phase (acute); (b) medium phase; (c) advanced phase.

In the acute post-stroke phase, functional improvements dependent on the stabilization of the brain injury, such as the reduction of edema, can be determined. Subsequently, neurological recovery occurs through a process of structural and functional reorganization of the central nervous system called neuroplasticity.

After stroke, multiple functional deficiencies are installed, and among these we present the following:

Motor impairments associated with lack of strength, balance, and coordination (paresis, paralysis); affections of superficial sensitivity (tactile and thermal with different degrees of hypoesthesia) with deep and mixed sensitivity; sensory, perceptual, and cognitive impairments and deficits; painful disorders from peripheral neurological disorders (neuralgia, neuritis) and central neurological disorders (thalamic pain); vegetative disorders; impaired osteotendinous reflexes in central motor neuron syndrome; dysarthia, dysphagia, aphagia.

Neuroscience studies have made clear progress, showing that the human brain is capable of plastic rehabilitation after injury [8].

The purpose of neurorehabilitation is to use the maximum capacity the quality of neuroplasticity, which is limited, but most people recovered reach a plateau of 70–80% of the initial post-stroke damage. Thus, as an uncomfortable recovery, compensatory movement strategies are applied, which help to alleviate motor deficiencies, for example, the use of a wheelchair [21].

The entire recovery of the motor function of a person with central nervous system injuries becomes a re-learning process, which uses the preserved sensorimotor circuits. The severity of lesions of the central nervous system and the individual neural capacity of the patient are the important aspects for re-learning and training of neural circuits in order to restore normal movement patterns.

5. Spinal cord injuries

Spinal injuries often cause serious injuries, due to their complexity, but also due to the neurological complications associated with them.

Thus, we mention that 40% of the injured people have cervical fractures, 15–20% have lumbar thoracic fractures, and in most cases, there are neurological lesions.

Statistically, trauma is caused by the following: 45% of road traffic accidents (car, motorcycle); accidents that occurred during work (falls from a height) in proportion of 20%; accidents in performance sports in the proportion of 16%; direct injuries caused by a firearm in the proportion of 15%; other types of trauma in the proportion of 4%.

In the case of people after the age of 75, about 60% of spinal fractures are due to falls from the same level below the evolution of osteoporosis.

Therefore, spinal cord injuries (SCIs) have an increased incidence, which lead to increased disability and lethality with difficulties in diagnosis and treatment by complex surgical procedures.

The percentage of vertebra-spinal cord injuries is, according to statistics, between 0.5 and 20% in the structure of the skeletal system and 10.5% in the structure of the nervous system.

The main location of spinal cord injuries is the thoraco-lumbar junction as shown by the following data: 12.3 of the injuries are located in the cervical area; 39.2% of injuries are located in the dorsal region; 48.5% of injuries are in the lumbar area; 25% T12 vertebrae for the dorsal region; 30% of the L1 vertebra in the lumbar region.

From a lethal point of view, the percentage reaches 34.4% for myeloma traumas, 8.3% for the dorsal region, 6% for lumbar traumas, most of which are with neuronal damage, with a disability of 95–98% of cases [19].

The mechanisms of trauma are generally by indirect mechanisms (95%), by the combined action of several forces, of different sizes, as follows: (1) Hyperflexion is the movement of excessive inclination in the anterior part of the transverse axis of the spine through the segment of movement. This movement performs a flexion, compression of the intervertebral bodies, and discs causing fractures by compression with the previous deformation of the vertebral body. The frequency of this type of trauma due to hyperflexion occurs in traffic accidents with frontal collision or by falls from a height; (2) hyperextension is the exaggerated tilt movement of the transverse axis of the segment in the back, common in the cervical vertebrae. Mechanism is caused by sudden rear-end traffic accidents and frontal collision accidents. Thus, this movement performed suddenly requires extension of the vertebral arches and produces compression, in unilateral or bilateral fractures of the bone elements; (3) pure compression is common in head-on sports accidents, head-on falls, or catapults in the cockpit. This axial compression of the anterior elements of the spine causes fractures of the vertebral body with centrifugal displacement of the fragments, especially when the spine is straight; (4) lateral hyperflexion puts pressure on the anteroposterior axis segment by making a forced frontal tilt either to the right or to the left. This movement is perfectly combined with rotation, in this case causing compression bone fractures (compression fractures) ligament injuries by traction; (5) shearing is obtained by the translational movement of parallel displacements of one vertebra over another, causing the destruction of the segment. The occurrence of this type of mechanism is especially in accidents involving the application of a ortho force on the spine (serial tamponade); (6) torsion is the twisting movement by rotating the motor segment, which causes dislocations of the intervertebral joints, but also severe sprains with spinal cord and joint complications.

Spinal cord injuries can cause a variety of pathologies, depending on the impact of the injury. Thus, the basis of these traumas is medullary or radicular neurological changes caused by symptoms, which overlap with the spinal shock (reversible loss of all spinal cord functions).

Injuries are classified as closed or open, with or without bleeding, with or without fractures, and stable or unstable.

The unstable injuries are dislocations between the spinous processes, dislocations of the vertebral arch, articular apophyses, vertebral blades, vertebral pedicles [22].

The effects of trauma and subsequent illness are as follows: Peripheral nerve palsy is a condition that can be acquired through a spinal cord injury or other causes such as diabetes, lead poisoning, alcohol. The symptoms are presented by muscle deformities, loss of muscle tone, lack of normal coordination of movements; paraparesis is an incomplete paralysis of the limbs due to spinal cord injury. It can be flaccid or spastic: vertebral fractures of vertebral bodies and arches. Stable fractures are those of the spinous, transverse processes, and vertebral body fractures, at which the settlement is below one third of the height of the vertebral body.

6. Methods and techniques of spine recovery

The methods and techniques of spine recovery are performed in specialized centers that offer sports and medical recovery programs for prophylactic, curative purposes, through physical, dynamic, and static exercises. The condition that the patient suffers from post-trauma, as well as its severity, orients the recovery plan toward muscle stimulation, improvement of muscular endurance, rehabilitation of diminished functions (balance, coordination, control, mobility).

From the point of view of post-trauma investigations, X-ray of the spine segment is performed to detect fractures, electrocardiogram for routine, tomography as an additional investigation, computed tomography for the skull, but also ultrasounds and Doppler for people with costal contusions.

The operative indication comes against the background of unstable fractures and medical urgency with the interest of the spinal cord.

From an orthopedic point of view, the treatment of thoracic spine injuries involves immobilization in the corset-lombostat and the small or large Minerva for cervical spine injuries. If the patient cannot be lifted and placed in the corset, a plaster bed mobilization is performed.

The initial goal in neuromotor recovery treatment is as follows: respiratory system; syndrome of vertebral insufficiency with painful symptoms, treated by physical therapy, massage, electrotherapy, and heliomarine cure; reduced urinary disorders by bladder re-education and acupuncture; motor deficits (paraparesis/folds, tetraparesis) with a re-education through physiotherapy and robotic equipment (Lokomat) with a variable duration of 3–5 years.

To ensure a complete recovery, treatment is essential in ensuring the natural healing processes. Thus, therapeutic methods include special techniques for soft and joint tissues, therapy with physiotherapy equipment, and pre- and post-treatment evaluation.

Immediate treatment of acute trauma involves reducing the inflammatory reaction by protection (immobilization in orthoses or plaster cast), rest, ice, compression, and lifting the limb above the horizontal plane.

The methods and techniques used to recover the spine are as follows: (a) Joint mobilization techniques are performed passively by the therapist or equipment, with the aim of restoring total mobility, and may be successfully combined with other soft tissue therapy and joint stabilization techniques; (b) essential muscle training techniques in the stagnation of muscle atrophy installed quickly post-trauma are applied depending on the type of trauma. Thus, the effort and the possibilities of loading are performed by increasing the speed of movement, endurance, the number of repetitions, or by changing the shape of the exercise. They can be performed simply under the guidance of a therapist or by using robotic infectious equipment or grounded platforms (Lokomat, Mit Manus); (c) strength training techniques; isometric, isotonic, isokinetic, with resistive band, with different resistances. Isometric training involves performing 5–10 maximum contractions per day, in order to protect the joints from stress and reduce inflammation. Isotonic training allows you to perform stretching exercises to prevent compressive forces on the affected joints. Training with bands of different resistance applied in multiple movement plans have an effect on stimulating the functions and actions specific to the area worked; (d) the manual techniques used in recovery are intended to relieve symptoms and improve the functioning of the spine. We can talk about the use of manual therapy techniques, massage, acupuncture, or presupposition, all of which treat the problems of the spine. Massage techniques are performed between the proximal to distal segments, with the aim of improving blood circulation, reducing inflammation and edema, but also to maintain the muscle tone of the surrounding areas. The application of manual therapy with addressability for the joint, muscular, and nervous areas is carried out with the aim of identifying the areas with joint pain or limitation and reducing them through manipulation, mobilization, or traction techniques. Acupuncture consists of applying a force of tension with the help of the fingers (thumb) on selected muscle areas for the purpose of stimulating muscle receptors, relieving symptomatic trigger points, and activating muscle tone by releasing substances necessary to reduce pain. Dry needling or acupuncture is the method of inactivating trigger points in chronic musculoskeletal syndromes. The method is performed by inserting needles perpendicular to the skin, which results in a muscular reflex response and increased post-treatment mobility; (e) proprioception and coordination re-education techniques represent the final phase of spine recovery, with the training of movements essential for the basic activity of the injured patient.

The benefits of using physical therapy in spinal recovery are as follows: Reduction of pain, stiffness of joints, and muscles; improving posture, balance, and coordination of the entire muscle and joint chain; recovery of acquired physical deficiencies after trauma (stroke, CVD); reintegration of patients into pre-trauma activities.

From the point of view of the technologies involved in the recovery of the spine and the neuromotor recovery of the spine, we mention the following: (i) training with treadmills with partial support and with the help of robots (Locomat, Armeos, Mit Manus); (ii) stimulating neuroplasticity using vibrating devices for the whole body (Galileo, Zeptor) and stimulating the healthy limb through Wii consoles; (iii) transcranial magnetic stimulation or direct currents with the neural helmet for the treatment of aphasia after stroke; (iv) electrotherapy with shock in reducing symptoms shows the use of ultrasound therapies, interference currents, shockwave, laser therapy, iontophoresis, as well as the combined technique.

Author details

Adrian Olaru
Robotic and Production Systems Department, University Politehnica of Bucharest, Industrial Engineering and Robotics Faculty, Robotics for Neurorehabilitation, Romania

*Address all correspondence to: aolaru_51@ymail.com

References

[1] Rinderu ET, Kinetoterapia in prezent. Necesitati, tendinte si posibilitati de integrare europeana. Bucuresti: Al V-lea Congres National de Kinetoterapie; 2005. p. 74

[2] Metzger JC, Lambercy O, Gassert R. Comparison of the performance of interaction control strategies on a hand rehabilitation robot. IEEE International Conference on Rehabilitation Robotics (ICORR). IEEE catalogue 846851. 2015. DOI: 10.1109/ICORR.2015.7281308

[3] Tamburrini G. Comunicarea dintre creier și computer. Perspective etice asupra modelelor de interacțiune. Neuroetică. 2009;**2**(3):137-149

[4] Khalili D, Zomlefer M. Un sistem robotic inteligent pentru reabilitarea articulațiilor și estimarea parametrilor segmentului corporal. IEEE Translations Biomedicine Engineering. 1988;**35**(2):138-146

[5] Baciu C. Chirurgia si protezarea aparatului locomotor. Bucuresti: Editura Medicala Bucuresti; 1986

[6] Dario P, Guglielmelli E, Laschi C. Humanoids and personal robots: Design and experiments. Journal of Robotic and Intelligent Systems. 2001;**18**(12):673-690

[7] Braddom R, et al. Physical Medicine and Rehabilitation. 4th edn. 2011; Chapter C5-6

[8] Balasubramanian S, Klein J, Burdet E. function. Current Opinion in Neurology. 2010;**23**(6):661-670

[9] Al B, Scurtu L. Anatomie si biomecanica. Bucuresti: Editura A.N. E.F.S; 1999;**25**

[10] Jacob BP, Gagner M. Robotică și chirurgie generală. Clinici chirurgicale din America de Nord. 2003;**83**(6):1405-1419

[11] Buch E, Weber C, Cohen LG, Braun C, Dimyan MA, Ard T, et al. Think to move: A neuromagnetic brain-computer interface (BCI) system for chronic stroke. Stroke. 2008;**39**(3): 910-917 (21-17)

[12] Carda S, Biasiucci A, Maesani A, Ionta S, Moncharmont J, Clarke S, et al. Electrically assisted movement therapy in chronic stroke patients with severe upper limb paresis: A pilot, single-blind, randomized crossover study. Archives of Physical Medicine and Rehabilitation. 2017;**98**(8):1628-1635 e1622

[13] Cipriani C, Zaccone F, Micera S, Carrozza MC. On the shared control of an EMG-controlled prosthetic hand. Analiza interacțiunii utilizator-proteză. Tranzacții IEEE pe robotică. 2008;**24**(1):170-184

[14] Dijkers MP, Bear PC, Erlandson RF, Kristy K, Geer DM, Nichols A. Patient and staff acceptance of robotic technology in occupational therapy: A pilot study. Journal of Rehabilitation Research and Development. 1991;**28**(2):33-44

[15] Kai-ming AA. Journal of Managerial Psychology. 2000;**15**(4):266-282. DOI: 10.1108/02683940010330957

[16] Kwakkel G, Kollen BJ, Krebs HI. Efectele terapiei asistate de robot asupra recuperării membrelor superioare după accident vascular cerebral: o revizuire sistematică. Reparare Neurorehabilation. 2008;**22**(2):111-121

[17] Nudo RJ, Plautz EJ, Frost SB. Rolul plasticității adaptive în recuperarea funcției după afectarea cortexului motor. Nervul Muscular. 2001;**24**(8):1000-1019 (13)

[18] Burgar CG, Lum PS, Shor PC, Machiel Van der Loos HF. Development of robots for rehabilitation therapy: The Palo alto VA/Stanford experience. Journal of Rehabilitation Research and Development. 2000;**37**(6):663-673

[19] Siciliano B, Sciavicco L, Villani L, Oriolo G. Robotică. Milano: Modelare, planificare și control; 2008. p. 3

[20] Dietz V, Sinkjaer T. Spastic movement disorder: Impaired reflex function and altered muscle mechanics. Lancet Neurology. 2007;**6**(8):725-733

[21] Dietz V, Fouad K. Restaurarea funcțiilor senzoriomotorii după leziunea măduvei spinării. Creier. 2014;**137**(Pt 3): 654-667

[22] Toscano J. Prevention of neurologic deterioration before admission to spinal cord injury unit. Paraplegia. 1988;**26**(3):143-115

Section 2

Artificial Muscles and Human Rehabilitation

Chapter 2

Characterization and Integration of Muscle Signals for the Control of an Exoskeleton of the Lower Limbs during Locomotor Activities

Jinan Charafeddine, Samer Alfayad, Adrian Olaru and Eric Dychus

Abstract

Daily activities are a source of fatigue and stress for people with lower extremity spasticity. The possible aids must be introduced while maintaining priority control by the patient. This work aims to develop such an application in the context of walking on the exoskeleton developed at the Systems Engineering Laboratory of Versailles (LISV). The application results are based on data recorded at the END-ICAP laboratory with gait sensors for healthy subjects, people with CPs, and people who had a stroke. Our contribution is the proposal of a new method of neuromotor control for a rehabilitative exoskeleton. It consists in determining and assisting the motor instructions for the movements of a patient while retaining his expertise; the assistance as needed and the detection of its intention based on a fusion of information. The results show that the proposed index characterizes the relationship of the angle difference with a reference movement for each joint. It dynamically compensates for movements efficiently and safely. This index is applicable for gait pathology studies and robotic gait assistance.

Keywords: gait pathology, muscle co-contraction, rehabilitation exoskeleton, neuro-motor control, patient expertise

1. Introduction

Our body movements are the result of a complex interaction between the central nervous system (CNS), nerves and muscles. Damage to any of these components can lead to movement disorders [1]. In the world, these lesions represent the leading cause of disability. A first example is that of cerebrovascular accident (stroke) (CVA). There are more than 700,000 new cases of stroke each year - one every four minutes. Stroke is the leading cause of acquired physical disability in adults, in cases where the patient survives. It can occur at any age, 25% of patients are under 65 and 10% under 45.

The number of cases in young people has increased significantly. A second example is cerebral palsy CP which affects a newborn every 6 hours or one in 800 children. What was previously called cerebral palsy is the leading cause of motor disability in children. These neurological conditions produce disorders that lead to slowed or absent voluntary movement caused by muscle spasticity. They affect the speed, quality, and ease of day-to-day activities of the human body [2], spasticity being an increase in muscle contractions that causes stiffness or contraction of muscles which impedes movement. It causes difficulty in walking, locomotion, or maintaining normal posture and balance [3].

As part of this work, we were interested in the design of rehabilitation systems for stroke and CP patients. The system considered is that of the rehabilitation exoskeleton, as it exists in clinics. The exoskeleton is a mechatronic device, worn by the patient, designed to increase physical performance, which is adapted to the shape of the human body and the function targeted for rehabilitation. This orthosis is used to provide high-intensity training to human limbs, tailored for each user, to promote recovery from a disease or neurological disorder [4]. The exoskeleton works mechanically in parallel with the human body [5] and can be controlled in a way passive or active. As a general rule, an exoskeleton considered as a whole is a system that has means of perception to acquire physiological signals of interest, of means of calculation to apply a treatment to extract the relevant parameters, and of means of action to give control instructions to mechanical effectors. In the case of walking disorders, the autonomic exoskeletons are frequently used for rehabilitation. This type of exoskeleton represents great challenges in terms of control: each patient has their own motor skills which make generic control of an exoskeleton difficult [6]. It is, therefore, necessary to find solutions that allow better control of the functions of these exoskeletons which place the patient in an expert position in his movement. This approach helps to promote rehabilitation to compensate for difficulties until almost normal movement is achieved. The patient retains the possibility of direct control by an interactive control method, between the exoskeleton and the patient, which takes into account the components of normal walking to establish a diagnosis and perform the analyzes necessary to understand walking pathological. The use of biomechanical parameters makes it possible to meet the medical needs concerning the gait of the patients to be treated [7]. First, the temporal-space parameters characterize walking overall, by integrating notions such as cadence, speed, or the number of steps [8]. Second, the kinematic parameters of gait are obtained by analyzing joint angles during a cycle by direct visual observation or by 3D video analysis for greater precision [9]. Finally, electromyographic (EMG) signals vary with time and can be characterized by their amplitude, frequency and phase [10].

The main challenge facing interactive exoskeletons is the direct relationship between biomechanical signals and the desired behavior of the exoskeleton. One of the most difficult cases addressed by robotics research is walking rehabilitation with an autonomous exoskeleton. The best-known autonomous systems are HAL [11], Wandercraft [12] or Ekso GT [13]. There are two known control strategies for these devices: (i) impedance or admittance control, generally predetermined and not taking into account the user's physical condition [14, 15], (ii) a control using electromyography by the detection of muscle activation, but this type of control is not satisfactory because the EMG signals recorded on patients with muscle disorders lead to incorrect operation [14–16]. Thus, these two strategies can only be used with a predefined behavior that the patient should follow. They are not suitable for patients with CP or after a stroke. Thus, these two target populations of our study require a new approach, based on continuous measurements of biomechanical signals.

The signals acquired must be analyzed, interpreted, and used to drive the trajectory of an online exoskeleton. Off-line approaches limit the patient's movement and do not take into account the variability of gait [9]. These cause fatigue and pain in the patient during a rehabilitation session with the exoskeleton [17]. Moreover, it does not allow to distinguish the deviations from the compensations due to the influence of the walking speed [9].

The approach we had followed is based on recording EMG signals from muscles involved in the knee and hip movements [18].

Like o novelty from the other research, we introduced a new muscle co-contraction index for a control strategy capable of assessing a compensatory joint angle suitable for CP or stroke patients. EMG signals contain information about the patient's intention [19] and can be used to assess muscle co-activation around the joint [20]. It has been suggested that muscle co-activation indicates the achievement of motor skills [21, 22] and it is also linked to joint stability [22]. Co-activation is considered to be an important factor contributing to the ineffectiveness of pathological movements [23]. Most researchers and clinicians rely on EMG measurements to express it as a comparison between the EMG measured for the muscles involved and the reference values. These values are also often examined using the co-contraction index [24] for a given joint. Although, the assessment of co-activation is suitable for off-line analysis and diagnosis, it does not meet the constraints of a control strategy because it is calculated at each phase of the walk cycle and in a specific way.

2. Theoretical framework

Locomotion is the ability to move. More specifically, this term refers to the way we travel from one place to another. Running, swimming, jumping, or flying are examples of types of locomotion. Human locomotion begins with signals from the central nervous system (CNS), which are transmitted to muscles to move joints. The CNS is made up of many nerve cells that are interconnected. Voluntary movement results from sending a nerve impulse from the brain that travels through the spinal cord and then the motor nerves. The brain plays a major role in the balance and coordination that are essential for our locomotion. The limbs are the main elements that drive movement in humans. The structures used during locomotion include the bones that support the body and help maintain the shape of the body and provide a surface for muscle attachment. Then there are the joints which are the points of contact between the bones and which allow the bones to move against each other without friction. Tendons connect muscle to bone and transfer the force generated by muscle contraction into the movement of the skeleton. Finally, the muscles that work in agonist/antagonist pairs move the bones of a joint by varying the angles of the joint. Locomotion is a behavior capable of providing information on motor control strategies [25–27]. Locomotor activity is described using specific anatomical terms that are used to determine the processes of movement. The terminology used describes this movement as a function of its direction with respect to the anatomical position of the joints as shown in **Figure 1a**. In general, the movement is named according to the anatomical plane in which it occurs (**Figure 1b**), where two axes of a joint are near or far from each other (**Figure 1c**). Anatomists use a unified set of terms to describe most movements, although other more specialized terms are needed to describe the uniqueness of movements such as those of the hands or feet [28].

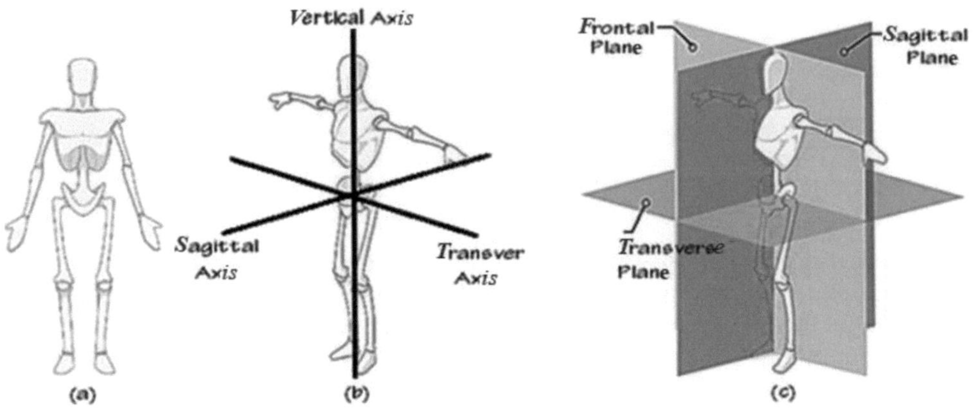

Figure 1.
Anatomical characterization of a movement: a) anatomical position; b) anatomical axes; c) anatomical planes.

A subject is considered to be in the anatomical position, shown in **Figure 1a**. When standing in an upright posture, face straight in front of him, feet close together and parallel, and palms of hands facing [29].

This is a standard position for which planes and axes are defined. To perform a practical analysis of human movement, anatomical movements can be defined as the act of moving body structures or changing the position of one or more joints of the body. Joint actions are described to the anatomical position which is the universal starting position for describing movement [30].

The movement of the limbs is depicted with the aid of reference planes, which can be seen in **Figure 1c**. From the reference anatomical position, three imaginary section planes are defined, crossing a subject's center of gravity and perpendicular to each other. These three planes are the sagittal plane, the frontal (or coronal) plane, and the transverse plane. The sagittal plane is a vertical plane perpendicular to the ground, which passes through the middle of the body and divides it into two symmetrical parts, lateral and medial or right and left. The frontal (or coronal) plane is a vertical plane perpendicular to the ground and to the sagittal plane, which separates the two anterior (front or ventral) and posterior (rear or dorsal) parts. The transverse plane, also called horizontal, axial, or median, is a horizontal plane parallel to the ground, separating the upper and lower parts of the body [31].

At the intersection of two planes constituting an axis, three anatomical axes are defined: the sagittal axis (anteroposterior), the transverse axis, and the vertical axis (longitudinal). The vertical axis is a longitudinal axis perpendicular to the ground when the subject is standing, the axis transverse is perpendicular to the vertical axis and the sagittal axis is an axis crossing the body from front to back and perpendicular to the previous two [31].

This work is particularly interested in pathological gait, and for this, it is necessary to understand normal gait and the terminology used to describe it. It is the benchmark against which a patient's gait can be compared. Even if the patient's gait differs from normal in one way or another, this does not necessarily imply a clinical or social need to transform it into a "normal" gait. Many abnormalities or walking disorders are compensation for certain problems encountered by the patient. Therefore, although abnormal they are useful.

A definition of walking is proposed by Smidt [32] as "The way of moving the body from one place to another by alternately and repetitively changing the position of the feet". In more detail, it is a translational progression of the body produced by coordinated rotational movements of the body segments. This essential activity, cyclical in nature, results from a series of rhythmic movements. It is characterized by alternating propulsive and restraining movements of the lower limbs which cause the center of gravity and the body to move forward by putting one foot in front of the other in a repetitive manner [33, 34]. Walking allows you to move around autonomously and independently. However, it is a complex activity, learned after a long process of trial and error before it matures. Once acquired, walking becomes an "automatic" activity that no longer requires special attention and is broken down into a series of movements that are repeated according to a specific cycle. The principles of gait are described in the literature of different disciplines: for example, from a medical point of view where the study of gait is carried out for surgery and prostheses [35] or else from the point of view of robotics, which aims at the design of artificial systems that move while walking [36].

Walking is accomplished using a complex, coordinated set of nerve signals, sent to muscles, to move joints, limbs, and the rest of the body. The central nervous system that produces these nerve impulse patterns is made up of neurons located in various parts of the brain and spinal cord.

Human locomotion relies on a rhythm-generating system located in the spinal cord and controlled by cortical neurons. This system reacts according to the information transmitted by the sensors of the muscles, joints, and skin of the legs [37]. At the cortical level, walking involves many structures such as the visual, cerebellar, vestibular, and proprioceptive cortices. In addition, the locomotor areas, trunk, and gray nuclei play a major role in controlling the activity of the gait spinal generator, which in turn controls muscle activity. The basal ganglia participate in the initiation of walking and its proper progress. The cortex plans the action and the choice of a motor program when the cerebellum is dedicated to controlling gait and balance, as shown in **Figure 1** [38].

The arthrology of walking is linked to the lower limbs, it is dependent on three main joints: hip, knee, and ankle. These joints are mainly stressed in one or two of the three planes defined in **Figure 1**. The directions of these movements for the hip, knee, and ankle are shown in **Figure 2**.

The possible movements of these joints, which take place in the sagittal plane, are flexion and extension, these same movements are also called dorsiflexion and plantarflexion for the ankle. Abduction and adduction movements take place in the frontal plane and are visible in **Figure 2**. Internal and external rotations (medial and lateral) take place in the transverse plane [31]. The muscles ensure the movement of the limbs by limiting the articular angles by their co-contraction. From a biomechanical point of view, muscles are distinguished by their role in mono or biarticulate movement, that is, by dividing the movements into anatomical units according to each joint. This distribution takes into account, due to the complexity of the musculoskeletal system, the fact that muscular action is not limited to a single degree of freedom mobilized voluntarily. According to Rasch et Burke [39] we distinguish: agonist muscles which are muscles whose contraction tends to cause the desired movement and antagonist muscles which are muscles whose contraction serves to produce an exactly opposite joint action. To fully understand normal walking, it is necessary to know which muscles are active as agonists and antagonists during different parts of the gait cycle to achieve the desired joint angles [13].

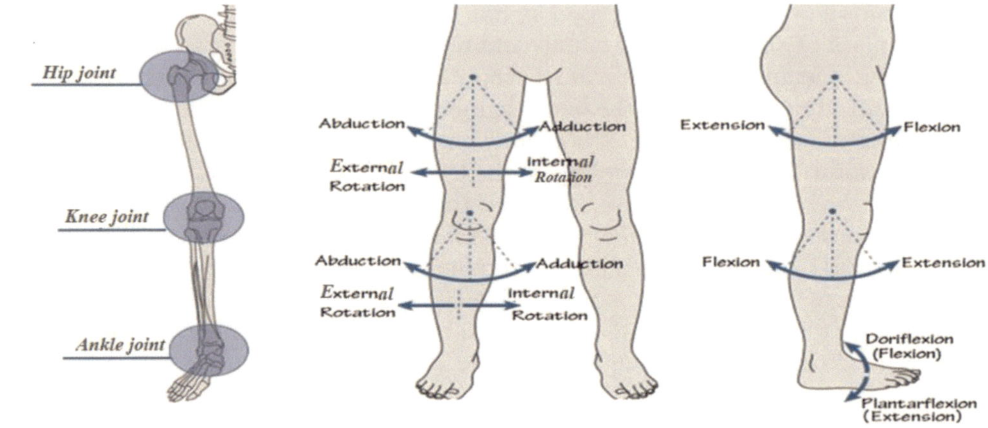

Figure 2.
Joints and main movements of the lower limbs.

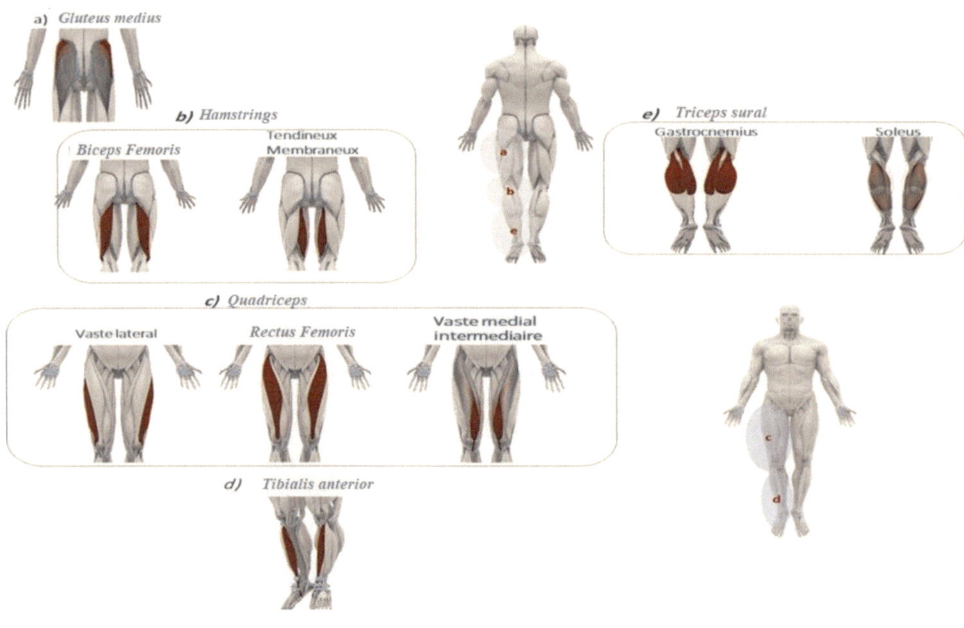

Figure 3.
The muscles: (a) gluteus medius, (b) hamstrings, (c) quadriceps, (d) tibialis anterior and (e) triceps sural.

For example, the action of the gluteus mediums, illustrated in **Figure 3a**, depends on a fixed point. If the fixed point is on the pelvis, the gluteus mediums abduct the thigh, with maximum effectiveness when the angle between the pelvis and the thigh is between 30° and 35°. The anterior fibers are brought into play as he rotates the thigh medially on the pelvis while the contractions of the posterior fibers cause the thigh to rotate sideways on the pelvis.

If the fixed point is on the femur, the gluteus medius muscle acts as a stabilizer of the pelvis. In addition, the gluteus medius performs a homolateral tilt of the pelvis. The hamstrings, visible in **Figure 3b**, are divided into semi-tendinous, semi-membranous, and biceps femoris. It causes knee flexion, thigh extension, pelvic retroversion. It participates in the stabilization of the knee and the pelvic girdle. In addition, they participate in the adduction of the thigh to the pelvis. During knee flexion, the biceps femoris rotates the knee sideways while the semi-tendon muscle rotates medially.

The quadriceps, visible in **Figure 3c**, is divided into 4 heads: the rectus femoris, the vastus medial, lateral, intermediate. Stabilization of the knee is affected by the vastus lateralis and medialis. The rectus femoris and vastus intermedius play a dynamic role, the rectus femoris muscle being a biarticular muscle, flexes the thigh on the pelvis. The tibialis anterior muscle is a flexor muscle where it exerts dorsiflexion of the ankle, this muscle is shown in **Figure 3d**. In addition, it causes an inversion of the ankle. Along with the contraction of other muscles, the tibialis anterior participates in triple flexion.

The sural triceps is divided into gastrocnemius medial, lateral gastrocnemius, and soleus **Figure 3e**. He exerts plantar flexion of the ankle which allows him to lift the whole weight of the body on the tip of the foot, as can be seen in **Figure 2**. There is little stress on the gastrocnemius during open chain knee flexion. The triceps sural participates with other muscles in knee flexion, but to a limited extent [Lacôte et al., 2014], its action is essential during the support phase for dorsiflexion control.

The role of muscles has been studied through electromyography (EMG), first by Scherb in 1940 who began his experiments by touching the muscles when walking a subject on a treadmill and then using the EMG. In 1981, the reference base for the study of muscle activity was made available by the group of "Verne Inman" at the University of California at San Francisco and at Berkeley. This group has contributed to significant advances in the understanding of muscle activity and many aspects of normal walking [40]. The use of surface EMG in gait analysis has since received much attention. We will detail two muscle groups that are particularly important for this work because they involve in the movements of the hip and the knee, which are the joints studied in this work. These are the quadriceps and the hamstrings. These two groups of muscles are biarticular and act on both the knee and the hip. At the front of the thigh, the quadriceps serves to straighten the leg and flex the thigh over the pelvis. At the rear, the hamstrings provide flexion and rotation of the knee as well as the extension of the thigh over the pelvis. The work is mainly carried out in the sagittal plane.

To take on the role of these muscle groups, it is important to study their action after neuromuscular activation. The kinematics is characterized in this plane by flexion of the hip and the knee. It is associated with median and posterior neuromuscular activations of the hamstring during the initial contact phase of the leg with the ground. The kinematics, in this plane, are also associated with an increased lateral neuromuscular activation of the quadriceps, during the maximum load phase of the movement [41]. We can thus see that:

- the rectus femoris (rectus femorus), which is part of the quadriceps, flexes the hip and allows the knee to extend;
- the biceps femoris, which is part of the hamstring, plays an opposite role to that of the rectus femoris and flexes the knee to allow hip extension;
- semimembranous and semitendinous, extend the hip and flex the knee.

The roles of these muscle groups are detailed in **Table 1** for each type of movement and each joint.

	Movement	Agonist	Antagonist
Knee	Flexion	Quadriceps	Hamstring
	Extension	Hamstring	Quadriceps
Hip	Flexion	Hamstring	Quadriceps
	Extension	Quadriceps	Hamstring

Table 1.
Agonist and antagonist bi-articular muscles, during hip and knee flexion and extension movements.

3. Biomechanics of walking

Biomechanics is a discipline that proposes to consider biological systems as objects of applied mechanics. His research methods are a combination of mechanics, anatomy, and physiology. It is a science of movement through the study and reproduction of the mechanisms which result in a definite movement of the body. Joint mechanics focus in particular on the joints, which provide both movements of a limb segment and its stability.

From the point of view of biomechanics, walking speed is decisive in evaluating the contribution of each segment of the body [42]. Walking speed primarily affects the stability and balance of the lower limbs. Joints can produce greater ranges of motion through greater muscle responses. In the bipedal system, the three main joints in the lower body and pelvis work in sync as muscles and impulses move the body forward. The degree to which the body's center of gravity shifts during translation defines the efficiency of the movement. The body's center of mass shifts sideways and up and down while walking. Walking is a repeating pattern comprising steps, with a stride denoting a complete cycle of walking. The characteristic frequency is defined by the split time, which measures the time elapsed between the heel touch of one leg and the heel touchdown of the contralateral leg. The width of a step can be described as the mediolateral space between the two feet [43].

Analysis of the gait cycle is important in examining biomechanical mobility to gain insight into lower limb dysfunction during dynamic movements [44]. By gait cycle analysis, it is best to examine each joint separately [42]. Both objective and subjective methods can be used [45, 46]. The cyclical nature of walking gives it a certain uniformity, but sometimes acyclic elements result from accidental causes such as tripping and other more natural causes such as spinning are severe. Since walking allows movement from one point to another but requires maintaining balance, dynamics play an important role **Figure 4**.

The automation of this activity facilitated its study, mainly by allowing biomechanical standards to be measured and established. The establishment of standards is essential for the study of pathological movement because they allow understanding by comparison. Walking is described in the state of the art as a cyclic function. A cycle is determined by all the events occurring between two successive identical events. The gait cycle is the time interval between two successive occurrences of one of the repetitive locomotion events [48]. The onset of the gait cycle is most often represented by the initial contact of a foot with the gait surface [13, 32, 47, 49, 50]. This cycle includes the support and oscillation phases. The oscillation phase refers to the moment when the foot is in the air for the progression of the limb. The support phase is the period when the foot is in contact with the ground as shown in **Figure 4**.

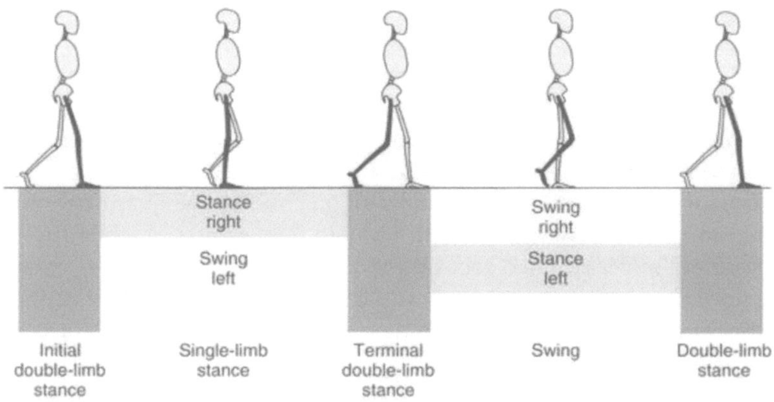

Figure 4.
The gait cycle and its phases [47].

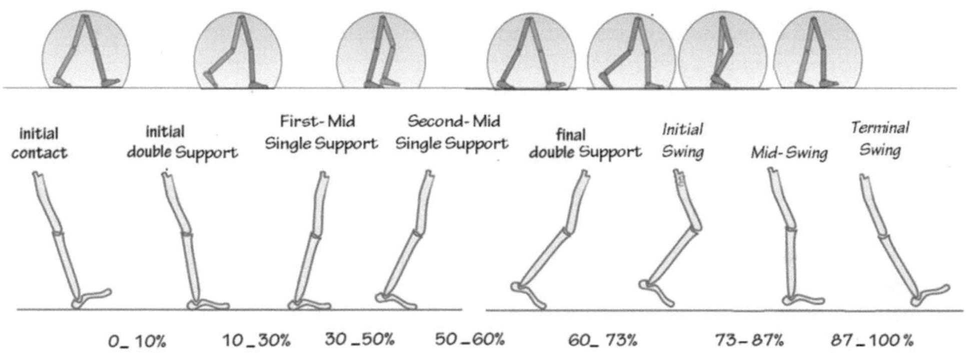

Figure 5.
Decomposition of the walking cycle.

Walking can also be defined as including initial double support, single-limb support, double end support, and a swing. The support phase represents 60% of the walk cycle, where each interval of double support is 10% and single limb support is 40% [50]. The oscillation phase designates the remaining 40%. Support from one limb is equivalent to tilting the other because they occur at the same time [47].

According to Perry and Davids [47], the walking cycle can be divided into eight phases, which correspond to three basic tasks as shown in **Figure 5**. (i) the first phase, from 0 to 2% of the cycle, begins with the initial contact or with the contact of the contralateral foot with the ground; (ii) the second phase, from 2 to 10% of the cycle, is the initial double push phase; (iii) the third phase, from 10 to 30% of the cycle, consists for the first half of the phase of simple support going from the detachment of the contralateral foot until the passing of the center of mass of the vertical; (iv) the fourth phase, from 30 to 50%, is the second half of the single support phase ending when the contralateral foot touches the ground; (v) the fifth phase, 50–60%, is the final double-support phase in which the weight is transferred from one limb to the other. The support phase ends with the toes peeling off; (vi) the sixth phase, from 60 to 73%, is the start of the oscillating phase and begins when the foot leaves the ground and ends when the foot is aligned with the contralateral foot; (vii) the seventh phase, 73–87%, is the mid oscillating phase, this phase ends when the oscillating member is forward; (viii) the eighth phase extends from 87 to 100% of the walk cycle, this is the end of the oscillating phase. This phase ends when the foot comes into contact with the ground. Advancement of the limb is complete when the leg segment is located in front of the thigh segment, all the steps are illustrated in **Figure 5**.

During walking, significant movements occur in the three sagittal, frontal and transverse planes. However, the study of gait movement for exoskeletal rehabilitation focuses on the sagittal plane. The kinematics of gait shows the quantification of joint angles in flexion and extension, **Figure 6**. The hip flexes and stretches once during the cycle. The flexion limit is reached around the middle of the first phase, then the hip is kept flexed until the first contact. Maximum extension is reached before the end of the support phase, and then the hip begins to flex again. The knee has two flexion peaks and two extensions during each walk cycle. It is nearly extended before initial contact, followed by flexion, narrowing towards extension during the downforce phase, before resuming flexing to peak in the swing phase. Finally, it stretches out again to prepare for the initial contact of the next step. The ankle is usually a few degrees from the neutral position on initial contact. After the initial contact, the plantarflexion control of the ankle brings the forefoot back to the ground. Halfway through, resumes dorsiflexion. Before the initial opposite contact, the angle of the ankle changes again, major plantarflexion occurs just after the toes have lifted off. During the swing phase, the ankle returns to dorsiflexion until the forefoot has cleared the ground, then the ankle returns to a neutral position which is maintained until initial contact, of the next step.

Muscle co-contraction (CCM) has a functional role in ensuring the stiffening of the joint to limit imbalance during limb advancement [51]. This role is particularly visible in children, in whom the establishment of this mechanism during the first steps has been well described [52, 53]. Gait patterns in people with a neurologic impairment are characterized by abnormal total voluntary contraction (TVC). This is reflected in particular by problems in postural stability [54], since it is important for providing joint stability [54–58], adequate movement precision and energy efficiency [59], as well as to adapt to environmental requirements [60]. From a neurological point of view, CCM is particularly important for rehabilitation, in robotics approaches to rehabilitation, it is even considered as a distinct element of motor control in many theoretical models of motor control [61–65].

TVC is a contraction of the antagonist's muscle, triggered by the command on the agonist, which creates an opposing or even reversing torque of the desired movement [66]. It characterizes the simultaneous activation of agonist and antagonist muscles within the same joint, and which act on the same plane [67].

TVC can be characterized physiologically and biomechanically. In this work, we will focus more particularly on bio-mechanical aspects. In the case of normal movement, the presence and degree of muscular co-contraction (CCM) are still subject to

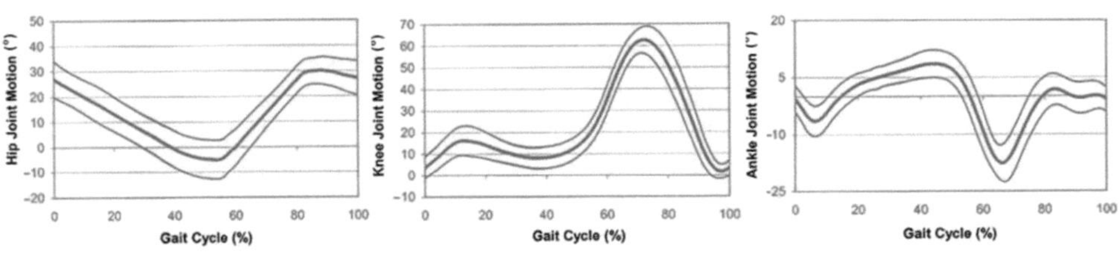

Figure 6.
Illustrations in the sagittal plane of the articular angles of the hip, knee and ankle, during a gait cycle performed by healthy adults at the spontaneous speed [adapted from (Lacôte et al., 2014)].

debate [68, 69]. The level of CCM depends on several parameters: it is proportional to the speed [70, 71], it is variable for the degree of inertia [72], to the muscle group under consideration [73] and maybe partially dependent on sex and age [74–76]. Other factors can also be taken into account to explain these variations [25, 53, 77–79] in the literature, but it is always calculated as a ratio between agonist and antagonist muscles and as a function of the walking sequence (double support, unipodal phase, etc.). From a scientific point of view, TVC has been defined in different ways to facilitate its interpretation. For walking, several components are thus put forward, such as magnitude, time or the temporal evolution of the amplitude. From these definitions, different formulas or computational approaches have been used to quantify the TVC [80]. All of these methods limit the comparison of data between studies and the understanding of TVC mechanisms.

The CCM assessment uses indices that have been created to contribute to a better understanding of the underlying mechanisms. These clues shed light on the role of TVC in walking in people with stroke or CP. During walking, TVC is presented as the time and magnitude of simultaneous contraction between opposing muscles [80]. The studies on the evaluation of TVC are purely descriptive, they do not involve any intervention or program of work with the patients. Only two experimental studies used control groups and assessed walking before and after an intervention [81–83].

TVC of agonist and antagonist muscles causing flexion/extension movement of a joint can be assessed in several ways using surface electromyography (EMG): (i) the first method is the visual estimate of the EMG amplitude or the percentage of overlap of the considered muscles; (ii) the second method aims to determine the relationship between the activity of the agonist and that of the antagonist; (iii) the third method quantifies the antagonistic moment by a mathematical model; (iv) the last method determines the ratio of the antagonist EMG to the EMG of the same muscle during its maximum agonist contraction. However, external factors can interfere with the measurements. For the first method, a standardization to allow comparison between subjects is not all time possible. For the second method, a decrease in agonist recruitment may introduce a bias to consider high co-contraction indices when they are only high by the decrease in agonist activity. This situation is all the more common as the quantitative values of these two EMGs are recorded under different conditions [66]. The hypothesis at the heart of the third method postulates a linear relationship between the EMG and the moment, the EMG is used as a surrogate for the force produced with all the inaccuracies that this can cause. The latter method has an advantage, which is that the antagonist is also measured in its role as an agonist. However, this method does not take into account the resulting joint moment, although it can be assessed separately.

The EMG thus plays a role in detecting the patient's intention [84], an intention that is considered to be a necessary parameter for the control mechanism of a rehabilitation exoskeleton. The objective of the work is to find a relationship between muscle co-contraction and joint angle to control a gait exoskeleton. This objective led us to use the last method, this choice is justified by a study of the indices of the proposed CCMs. Among the different methods of calculating the CCM index, three are more robust and more precise. In addition, these methods are evaluated in cases of neurological diseases, such as those studied in this work. These CCM indices are shown below:

(i) the first modified method that was introduced by Unnithan et al. [77] and by Frost et al., [53]:

$$\mathrm{ICCM}_1(t) = \frac{surface\ commune}{surface\ totale} \times 100 \tag{1}$$

where the common area is the integral of the sum of the antagonistic and agonistic EMG amplitudes and the total area is the total number of data points;
(ii) the second method has been proposed by [81]:

$$\mathrm{ICCM}_2(t) = \frac{2 \times surface\ commune}{surface\ EMG_{antago} + surface\ EMG_{ago}} \times 100 \tag{2}$$

where the EMG surface is the integral of the normalized muscle EMG values.
(iii) the third method is the one defined by Falconer and Winter [85]:

$$\mathrm{ICCM}_3(t) = \frac{2 \times \left\| EMG(t)_{antago} \right\|}{\left\| EMG(t)_{antago} \right\| + \left\| EMG(t)_{ago} \right\|} \times 100 \tag{3}$$

Note that these CCM indices are estimated in the state of the art as values like values discrete for each sub-phase of a walking cycle.

4. Experimental research

The experimental research has the goal to define one new form of the CCM indicators (4, 5, 6) and neuro motor indices (INM). The precise determination of neurological impairment in terms of CCM during gait requires robust measurement techniques that take into account the environmental conditions under which gait is assessed [86]. For example, walking on a floor surface rather than on a treadmill, walking at different speeds and for longer distances or times increase TVC recruitment and variability between subjects [87, 88]. TVC during walking has been studied in patients with stroke [54, 81, 89–93], and in the case of CP patients [77, 81, 94–99].

Bio kinematic analysis relies on two-measure detection: (i) first, the angles of the knee and hip in the sagittal plane when flexing and extending these joints when walking. From these angles, a percentage interpolation is estimated over a full gait cycle; (ii) next, EMGs, designed to assess TVC during functional movement. Their study requires an analysis of the relative variations in contraction over time between the agonist and the antagonist [100].

There are standards that have been developed for the different stages of processing of these signals, such as SENIAM for analog and digital acquisition and analysis [101], their implementation however remains variable. Concerning the most appropriate techniques for analyzing EMG signals, differences remain as for the choice of the normalization technique, which leads to important differences between the studies [102].

I have set up the following chain to process the EMG data. EMG measurements are filtered with a 4th order bandpass (10-400 Hz) Butterworth filter, centered to eliminate their average, then rectified by calculating their absolute value, then filtered by a low-pass filter with a frequency f_c between 4 and 6 Hz, the value of which depends on the cadence of the subject [103].

The first processing step is to interpolate the joint angles, taking care to generate the same number of points as those of the EMG matrices. This makes it easier to align treatments later. It is then important to estimate the mean and the variance over several walking cycles to obtain an average curve that takes into account the variation margins for hip and knee angels, **Figure 7**.

The second step is to extract the TVC indices, to obtain continuous values. The goal is to have the data necessary to establish a relationship between joint angles and TVC indices during the gait cycle. It is thus possible to use a correlation method between each TVC index and the angular variation curve for each joint. This analysis is done for each part of the walking cycle in healthy subjects **Figure 8**.

The femoral quadriceps is the largest muscle group in the human body. It is he who mainly supports the weight of the body and allows movement. It performs a role of knee extensor and hip flexor via two muscles included in our study: the rectus femoris (RF) and the vastus lateralis (VL), described below. The hamstring is a muscle group that allows hip extension and knee flexion by two muscles, the biceps femoris (BF) and the semi membranous (SM). From the perspective of a system using the minimum number of electrodes, we chose to measure the LF and RF. This is because the BF is a very large muscle, which extends from the lower pelvis above the knee. It is particularly involved in the flexion of the leg at the knee and in the extension of the thigh at the hip, playing a regulating role in the flexion of the knee during walking. The DF is also very extensive and also plays a role in stabilizing the knee during walking, as shown in **Figure 9**. The VL allows the extension of the leg at the knee and secondarily the external rotation of the leg but does not play a role in the flexion of the hip. SM is a thigh extensor muscle that flexes the leg on the thigh, but when the knee is flexed the SM acts as an internal rotator of the leg. The VL and the MS can therefore be left out for this study.

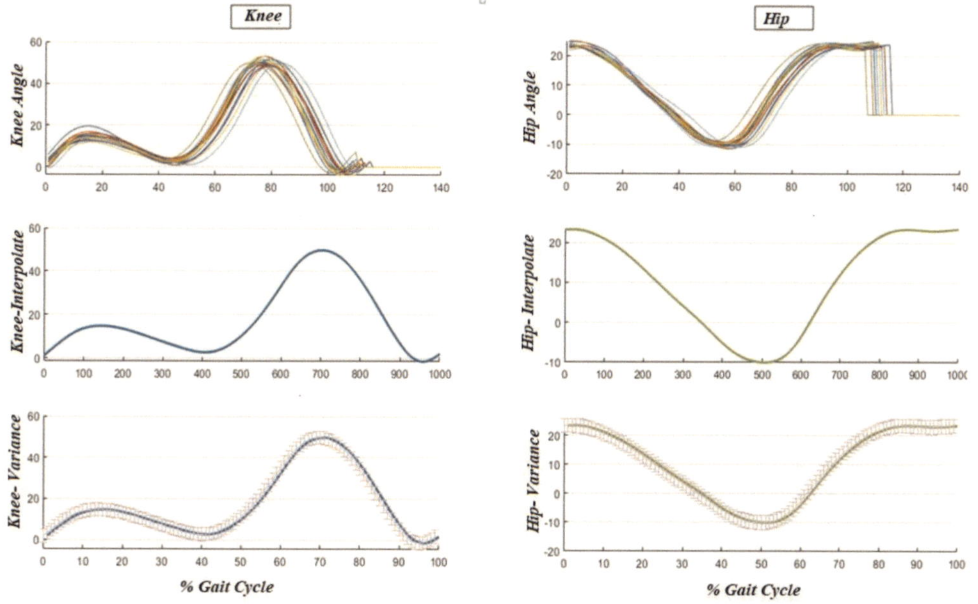

Figure 7.
Mean and variance of angular curves of the knee and hip of a healthy subject.

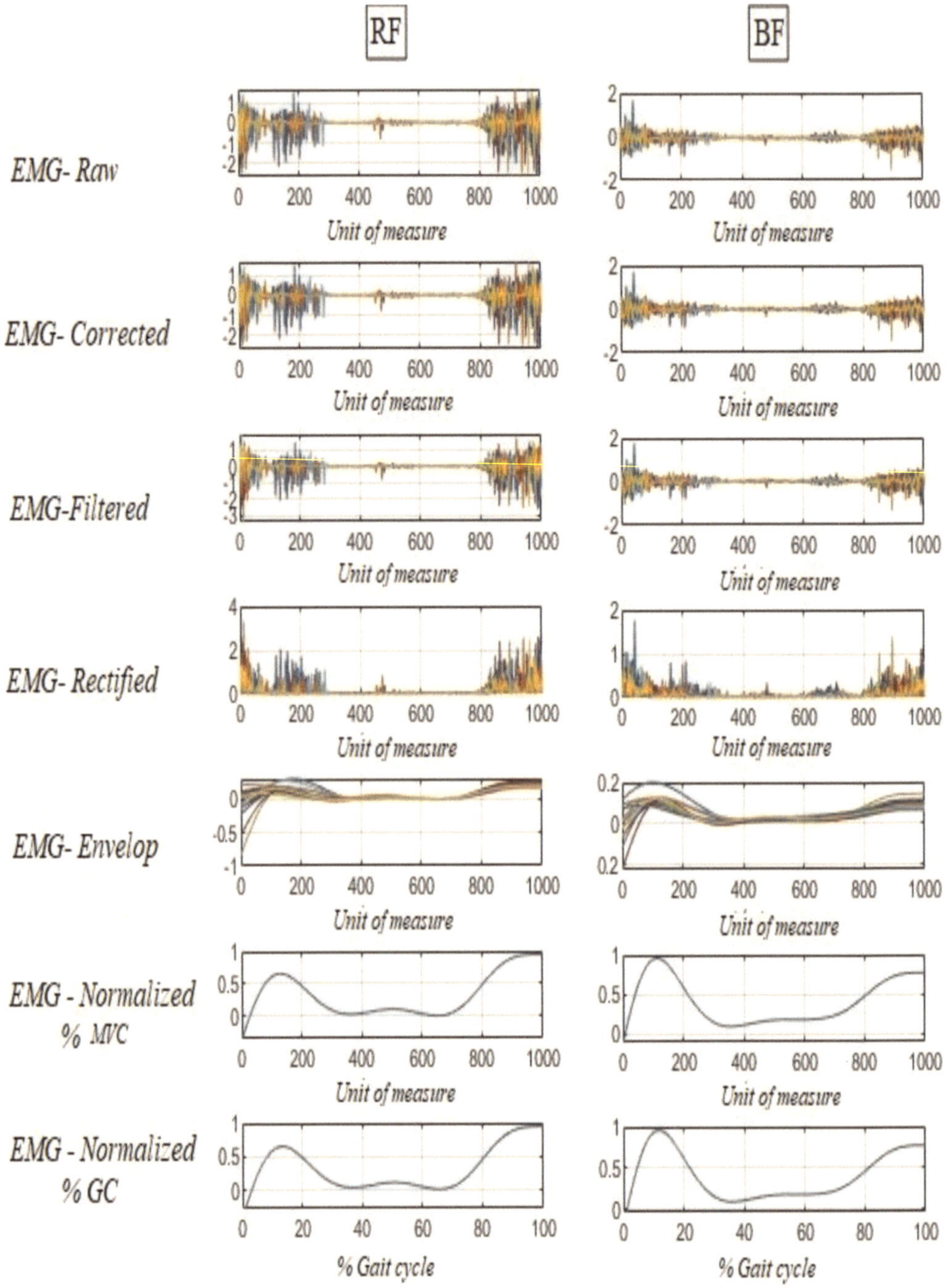

Figure 8.
Example of processing of RF - quadriceps EMG signals and BF- hamstring EMG signals for a healthy subject during spontaneous walking at normal speed.

The CCM indices estimated in this work follow the methods described in the state of the art. In the reference articles, they are assessed in a discretized fashion for each part of the walking cycle. This type of approach does not allow continuous estimation,

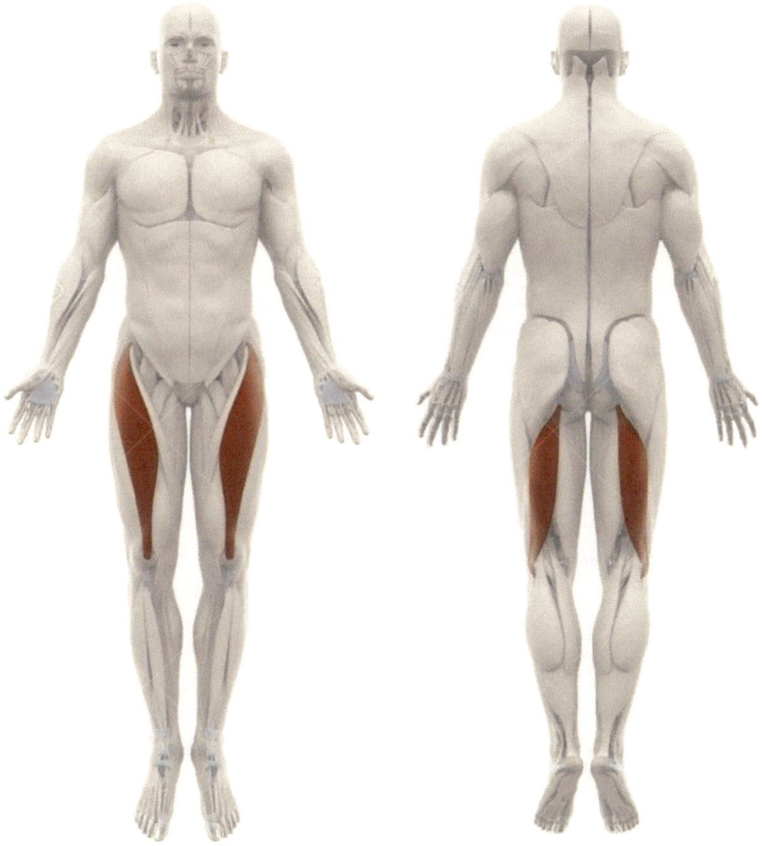

Figure 9.
Biceps Femoris (BF) and rectus Femoris (RF).

which is necessary to complete this work. We proposed a reformulation of Eqs. 1–3 which allows these TVC indices to be determined continuously, using a sliding window with overlap. The proposed reformulation of the TVC indices is shown below and is used for the estimates shown in **Figure 10**.

The method of Unnithan [76] and of Frost et al [53] (1) is thus rewritten:

$$\mathrm{ICCM}_1(t) = \frac{\left(\int_{t_1}^{t_2} \mathrm{ENV}_{\mathrm{ago}}^{\mathrm{emg}}(t) \cap \mathrm{ENV}_{\mathrm{anta}}^{\mathrm{emg}}(t)dt\right)}{\int_{t_1}^{t_2} \mathrm{ENV}_{\mathrm{ago}}^{\mathrm{emg}}(t) \cup \mathrm{ENV}_{\mathrm{anta}}^{\mathrm{emg}}(t)dt} \times 100 \tag{4}$$

The method of Hesse et al [81] (2) becomes:

$$\mathrm{ICCM}_2(t) = \frac{2\left(\int_{t_1}^{t_2} \mathrm{ENV}_{\mathrm{ago}}^{\mathrm{emg}}(t) \cap \mathrm{ENV}_{\mathrm{anta}}^{\mathrm{emg}}(t)dt\right)}{\int_{t_1}^{t_2} \mathrm{ENV}_{\mathrm{ago}}^{\mathrm{emg}}(t) + \mathrm{ENV}_{\mathrm{anta}}^{\mathrm{emg}}(t)dt} \times 100 \tag{5}$$

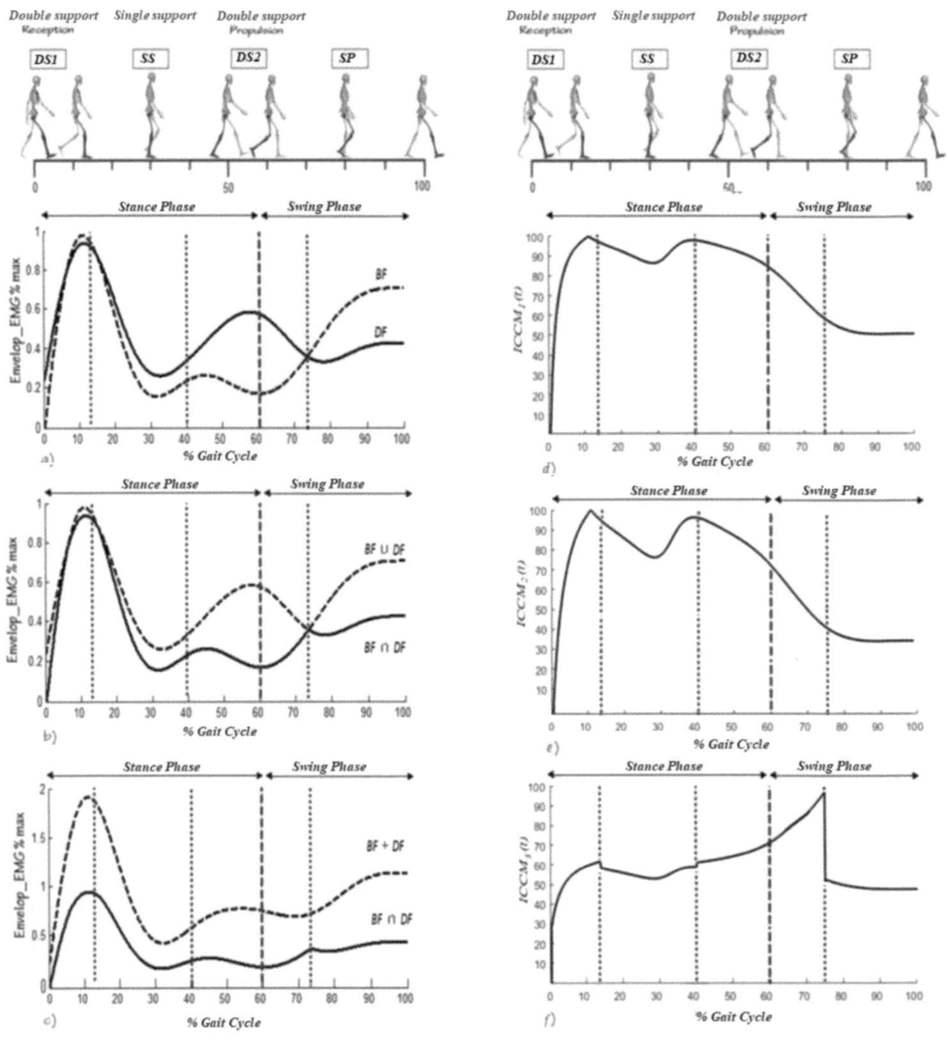

Figure 10.
Continuous assessment of co-contraction indices.

Finally, the method of Falconer and Winter [85] (3) can be reformulated as follows:

$$\mathrm{ICCM}_3(t) = \frac{2\left(\int_{t_1}^{t_2} \mathrm{ENV}_{\mathrm{ago}}^{\mathrm{emg}}(t)dt + \int_{t_2}^{t_3} \mathrm{ENV}_{\mathrm{anta}}^{\mathrm{emg}}(t)dt\right)}{\int_{t_1}^{t_3} \mathrm{ENV}_{\mathrm{ago}}^{\mathrm{emg}}(t) + \mathrm{ENV}_{\mathrm{anta}}^{\mathrm{emg}}(t)dt} \times 100 \tag{6}$$

For the study of the knee, $\mathrm{ENV}^{\mathrm{emg}}{}_{\mathrm{anta}}$ is the envelope of the muscles that train the slack knee flexion (quadriceps) and $\mathrm{ENV}^{\mathrm{emg}}{}_{\mathrm{ago}}$ is the envelope of the muscles that are train the knee extension movement (hamstrings). For the study of the soft-hip, $\mathrm{ENV}^{\mathrm{emg}}{}_{\mathrm{anta}}$ is the envelope of the muscles that causes movement flexion of the hip (hamstring) and $\mathrm{ENV}^{\mathrm{emg}}{}_{\mathrm{ago}}$ is the envelope of the muscles that evokes the hip extension movement (quadriceps). For the $ICCM_1(t)$ and $ICCM_2(t)$ indices, the period of t_1 and t_2 represents a complete cycle of walking while for $ICCM_3(t)$, the period of t_1 to t_2 denotes the period when the agonist EMG is lower than the antagonist EMG, whereas from t_2 to t_3, they denote the period when the antagonist EMG is lower than the agonist EMG.

Compared to **Figure 10**, $ICCM_3$ shows a discontinuity at the start of each step cycle sub-phase. The other two indices $ICCM_1$ and $ICCM_2$ are continuous variables relative to the walking cycle. So, in the rest of this work, we will leave $ICCM_3$ aside. We examined the relation between ICCM selected and joint angles, the result was, no relation, especially in extension (**Figures 11** and **12**).

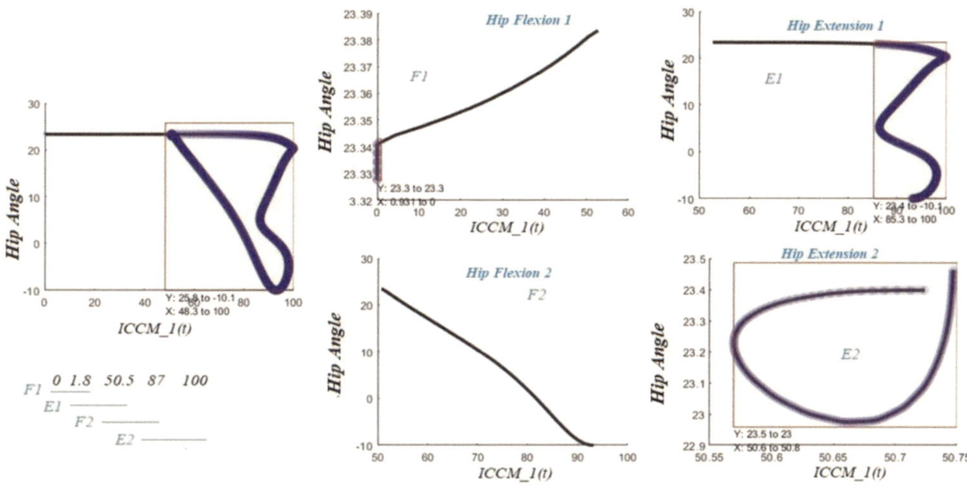

Figure 11.
Decomposition according to hip flexion and extension θ_{hip} as a function of $ICCM_1$.

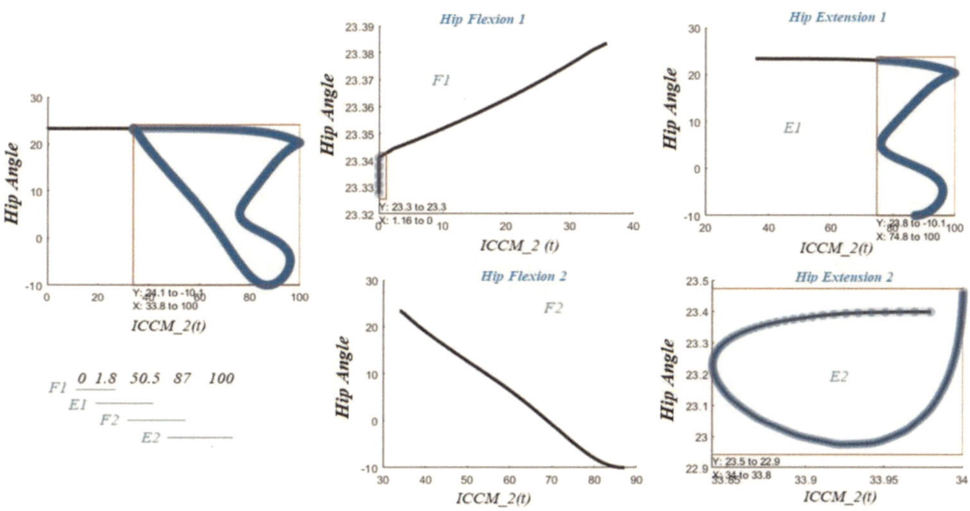

Figure 12.
Decomposition according to hip flexion and extension θ_{hip} as a function of $ICCM_2$.

This research paper was introduced one new movement indicator- Neuro Motor Indices (NMI). The NMI is by construction defined as an index of continuous muscle co-contraction and directly dependent on the trajectory of the joint angles. Our proposal is to combine the two chosen CCM indices, $ICCM_1$ and $ICCM_2$, to gain the advantages of each, using a relation that depends on the flexion/extension of each joint during a movement. This approach is inspired by the flexion/extension of each joint during the gait cycle. The NMI is thus built as a relation between envelopes of an agonist/antagonist muscle pair according to this flexion/extension. In practice, the combination of TVC indices is performed with a nonlinear regression derived from a polynomic Hermitian. Thus, INM is a non-linear combination of $ICCM_1$ and $ICCM_2$. TVC indices are defined as indices of joint stability and are calculated from EMG signals that detect the subject's intention. The INM, therefore, takes advantage of these properties and defines a new relationship between EMGs and joint angles, which depends on the capacity of the muscles and therefore on the expertise of the patient.

The formal definition of INM is as follows:

$$\mathrm{A}(t) = \frac{2\left(\int_{t_1}^{t_2} \mathrm{ENV}_{\mathrm{ago}}^{\mathrm{emg}}(t) \cap \mathrm{ENV}_{\mathrm{anta}}^{\mathrm{emg}}(t)dt\right)}{\int_{t_1}^{t_2} \mathrm{ENV}_{\mathrm{ago}}^{\mathrm{emg}}(t) + \mathrm{ENV}_{\mathrm{anta}}^{\mathrm{emg}}(t)dt} \times 100$$

$$\mathrm{B}(t) = \frac{2\left(\int_{t_1}^{t_2} \mathrm{ENV}_{\mathrm{ago}}^{\mathrm{emg}}(t) \cap \mathrm{ENV}_{\mathrm{anta}}^{\mathrm{emg}}(t)dt\right)}{\int_{t_1}^{t_2} \mathrm{ENV}_{\mathrm{ago}}^{\mathrm{emg}}(t) \cup \mathrm{ENV}_{\mathrm{anta}}^{\mathrm{emg}}(t)dt} \times 100 \tag{7}$$

$$INM = \frac{1}{2}\{\mathrm{A}(t) + R_x(t)\mathrm{B}(t)\}$$

$$R_x(t) = h_1(t)f_0 + h_2(t)p_0 + h_3(t)f_1 + h_4(t)fp_1,$$
$$f(t) = \mathrm{ENV}_{\mathrm{ant}}^{\mathrm{emg}}(t) \cap \mathrm{ENV}_{\mathrm{ago}}^{\mathrm{emg}}(t)$$

for which h_1, h_2, h_3, h4 ∈ P are the roots of a Hermitian polynomial, and p_0, p_1 are the tangents to f_0 and f_1, and (t_2 - t_1) represents a complete operating cycle. To test the reliability of the INM, canonical correlation analysis is applied in an offline study. Then, to move on to the control stage, a second correlation analysis is applied in an online study.

The results obtained from the analyzes of the estimate of error for the 20 subjects show that the healthy subjects (Example right hip of a healthy subject **Table 2**), that is to say, those who have a normal gait cycle, show a small change for all ICCMs with a

Estimation	DS1	SS	DS2	SP
$\epsilon ICCM_1 = f(\theta_{\mathrm{hip}}(t))$	2.3	6.7	5.3	7.2
$\epsilon ICCM_2 = f(\theta_{\mathrm{hip}}(t))$	4.4	7.1	5.1	6.9
$\epsilon ICCM_3 = f(\theta_{\mathrm{hip}}(t))$	7.5	8.1	6.1	7.9
$\epsilon INM = f(\theta_{\mathrm{hip}}(t))$	1	0.9	0.7	0.5

Table 2.
Comparison of the error estimate for the indices studied for the case of the right hip of a healthy subject.

Estimation	DA_1	SA	DA_2	PO
$\epsilon ICCM_1 = f(\theta_{\text{hanche}}(t))$	3.1	42.9	39.5	14.2
$\epsilon ICCM_2 = f(\theta_{\text{hanche}}(t))$	3	54.9	49.8	20
$\epsilon ICCM_3 = f(\theta_{\text{hanche}}(t))$	44.1	28.8	66.9	49.8
$\epsilon INM = f(\theta_{\text{hanche}}(t))$	2	1.9	1.8	0.9

Table 3.
Error estimates for the indices studied in the case of the right hip of a stroke subject.

slight advantage for the IMN. In the case of patients, the IMN provides valuable information and allows the evaluation of walks that deviate from prototypical cycles, as is most often the case in stroke or CP patients (Example right hip of a stroke subject **Table 3**). Analysis for neurological subjects shows that ICCMs are not appropriate. This calls into question their reliability in analyzing these cases, even though they are used in the state of the art. The cycles in stroke or CP subjects are poorly determined and they show very significant changes compared to normal walking cycles. However, they make it possible to achieve movement that allows movement and most often minimizes pain for patients. Thus, ICCMs worked well in subjects with normal walking, but the main source of error was in subjects with stroke or cerebral paralysis CP.

5. Conclusion

The importance of the use of mechatronic systems, especially exoskeletons, reveals their role as a solution for the rehabilitation of walking in people with neurological problems or who are forced to stay in a wheelchair. This work is primarily concerned with the possibility of implementing effective rehabilitation using these systems. This requires close human-machine collaboration between the patient and the exoskeleton. The central place of the patient's expertise in his movement and his ability to interact with the machine in the proposed work constitutes our main contribution to this research work. In this work, we, therefore, integrated the expertise of the patient into a controller that can be used on a rehabilitation exoskeleton. This model takes into account the effects of muscle weakness and spasticity. It is possible to use these results for a simple control strategy that is easy to evolve with the patient's condition. This type of strategy adapts to all types of gaits and all speeds of movement. The gait of an evolving exoskeleton of the lower limbs is treated in a work by the team of Samer AlFayed [104]. This exoskeleton is designed with two active degrees of freedom on the hip and knee. We presented a state of the art on biomechanics of gait, the existing methods for the control of the exoskeletons of the lower limbs used in the world, the neurological problems causing gait disturbances such as stroke and the CP.

The basic premise of this work is that if a movement involves EMG activity, we can then estimate the movement from the EMGs to respect the muscle capacity and will of the patient. The research question addressed here, therefore, is to find a relationship between bio- signals (EMG) and kinematic parameters (here articular angles) to drive an assistance exoskeleton.

The solution that we have proposed is based on the detection of muscle co-contraction between agonist/antagonist muscles. Co-contraction plays a crucial role in detecting patient intentions and in characterizing joint angles when walking. This co-contraction as well as the patient's intention are estimated from EMG measurements. We have defined the muscle function that allows us to estimate muscle co-contraction. We presented its evaluation in the state of the art from the indices of muscle co-contractions, and determined these indices in a continuous fashion between a pair of biarticular agonist/antagonist muscles for the hip and the knee. We then tested the correlation of these indices with joint angles and showed that the correlations are weak. This shortcoming is all the more limiting as the same co-contraction index is used in practice to assess two different joint angles. We, therefore, sought to introduce the specificity of each joint to these indices, by proposing to use a nonlinear regression related to the flexion/extension of each joint. We were thus able to propose a specific neuromotor index for each joint. We determined the error estimate for the INM and other state-of-the-art indices in the gait cycle sub-phases for healthy subjects, stroke, and CP. The INM showed better results than all other indices. To more fully validate this clue, we used a two-step canonical correlation analysis throughout the walk cycle. This evaluation showed the advantage of the INM over state-of-the-art indices. The INM shows a quasi-linear correlation with joint angles and the internal correlation was validated. In the future work will be comparatively studied all these indicators in different stages of the patient health and for all joints.

Author details

Jinan Charafeddine[1], Samer Alfayad[2], Adrian Olaru[3]* and Eric Dychus[4]

1 Paris Saclay University, France

2 Université d'Evry Val d'Essonne, Paris, France

3 University Politehnica of Bucharest, Romania

4 SANDYC Industries, Paris, France

*Address all correspondence to: aolaru_51@ymail.com

References

[1] Lee RJG, Tatton WG. Motor responses to sudden limb displacements in primates with specific CNS lesions and in human patients with motor system disorders. Canadian Journal of Neurological Sciences. 1975;**2**(3):285-293

[2] Brudny J, Korein J, Levidow L, Grynbaum BB, Liberman A, Friedmann LW. Sensory feedback therapy as a modality of treatment in central nervous system disorders of voluntary movement. Neurology. 1974;**24**(10):925-925

[3] Burne JA, Carleton VL, O'dwyer NJ. The spasticity paradox: Movement disorder or disorder of resting limbs? Journal of Neurology, Neurosurgery & Psychiatry. 2005;**76**(1):47-54

[4] Low KH. Robot-assisted gait rehabilitation: From exoskeletons to gait systems. In: 2011 Defense Science Research Conference and Expo (DSR). IEEE; 2011. pp. 1-10

[5] Pons JL. Wearable robots: biomechatronic exoskeletons. John Wiley & Sons; 2008

[6] Norouzi-Gheidari N, Archambault PS, Fung J. Effects of robotassisted therapy on stroke rehabilitation in upper limbs: Systematic review and meta-analysis of the literature. Journal of Rehabilitation Research & Development. 2012;**49**(4):479-496

[7] Isakov E, Mizrahi J, Najenson T. Biomechanical and physiological evaluation of FESactivated paraplegic patients. Journal of Rehabilitation Research & Development. 1986;**23**(3):9-19

[8] König N, Singh NB, Von Beckerath J, Janke L, Taylor WR. Is gait variability reliable? An assessment of spatio-temporal parameters of gait variability during continuous overground walking. Gait & Posture. 2014;**39**(1):615-617

[9] Simon SR. Quantification of human motion: gait analysis—benefits and limitations to its application to clinical problems. Journal of Biomechanics. 2004;**37**(12):1869-1880

[10] Haig AJ. Technology assessment: the use of surface EMG in the diagnosis and treatment of nerve and muscle disorders. Muscle Nerve. 1996;**19**:392-395

[11] Sankai Y. HAL: hybrid assistive limb based on cybernics. In: Robotics Research. Berlin, Heidelberg: Springer; 2010. pp. 25-34

[12] Gurriet T, Finet S, Boeris G, Duburcq A, Hereid A, Harib O, et al., editors. Towards restoring locomotion for paraplegics: Realizing dynamically stable walking on exoskeletons. In: 2018 IEEE International Conference on Robotics and Automation (ICRA). IEEE; 2018. pp. 2804-2811

[13] Miller LE, Zimmermann AK, Herbert WG. Clinical effectiveness and safety of powered exoskeleton-assisted walking in patients with spinal cord injury: Systematic review with metaanalysis. Medical Devices (Auckland, NZ). 2016;**9**:455

[14] Chen G, Chan CK, Guo Z, Yu H. A review of lower extremity assistive robotic exoskeletons in rehabilitation therapy. Critical Reviews in Biomedical Engineering. 2013;**41**(4-5):343-363

[15] Tingfang Yan, Marco Cempini, Calogero Maria Oddo, and Nicola Vitiello. Review of assistive strategies in powered lower-limb orthoses and exoskeletons. Robotics and Autonomous Systems, 64 :120–136, 2015.

[16] Steele C. Applications of emg in clinical and sports medicine. BoD–Books on Demand. 2012

[17] Wei Hong Y, King Y, Yeo W, Ting C, Chuah Y, J. Lee, and Eu-Tjin Chok. Lower extremity exoskeleton: review and challenges surrounding the technology and its role in rehabilitation of lower limbs. Australian Journal of Basic and Applied Sciences. 2013;**7**(7): 520-524

[18] Snyder-Mackler L, Ladin Z, Schepsis AA, Young JC. Electrical stimulation of the thigh muscles after reconstruction of the anterior cruciate ligament : effects of electrically elicited contraction of the quadriceps femoris and hamstring muscles on gait and on strength of the thigh muscles. Clinical Journal of Sport Medicine. 1992;**2**(3):227

[19] Lelas JL, Merriman GJ, Riley PO, Kerrigan DC. Predicting peak kinematic and kinetic parameters from gait speed. Gait & Posture. 2003;**17**(2):106-112

[20] Kellis E. Quantification of quadriceps and hamstring antagonist activity. Sports medicine. 1998;**25**(1): 37-62

[21] Le P, Best TM, Khan SN, Mendel E, Marras WS. A review of methods to assess coactivation in the spine. Journal of electromyography and kinesiology. 2017;**32**:51-60

[22] Basmajian JV. Motor learning and control : a working hypothesis. Archives of physical medicine and rehabilitation. 1977;**58**(1):38-41

[23] Winter DA. Biomechanics and motor control of human movement. John Wiley & Sons; 2009

[24] Rosa MCN, Marques A, Demain S, Metcalf CD, Rodrigues J. Methodologies to assess muscle co-contraction during gait in people with neurological impairment–a systematic literature review. Journal of Electromyography and Kinesiology. 2014;**24**(2):179-191

[25] Seger JY, Thorstensson A. Muscle strength and myoelectric activity in prepubertal and adult males and females. European journal of applied physiology and occupational physiology. 1994; **69**(1):81-87

[26] Patla AE, Calvert TW, Stein RB. Model of a pattern generator for locomotion in mammals. American Journal of Physiology-Regulatory, Integrative and Comparative Physiology. 1985;**248**(4):R484-R494

[27] Wootten ME, Kadaba MP, Cochran GVB. Dynamic electromyography. II. Normal patterns during gait. Journal of Orthopaedic Research. 1990;**8**(2): 259-265

[28] Burnfield M. Gait analysis: normal and pathological function. Journal of Sports Science and Medicine. 2010; **9**(2):353

[29] Palastanga N, Field D, Soames R. Anatomy and human movement: structure and function, volume 20056. Elsevier Health Sciences. 2006

[30] Marieb EN, Hoehn K. Human anatomy & physiology. Pearson. Education. 2007

[31] Michael W. Gait analysis an introduction. Oxford Orthopaedic Engineering Center: Elsevier; 2007

[32] Smidt GL. Gait in Rehabilitation. Churchill Livingstone; 1990

[33] Delaney R. Measuring walking: A handbook of clinical gait analysis. British

Journal of Occupational Therapy. 2014; **77**(5):264-265

[34] Norkin CC, Levangie PK. Joint Structure & Function: A Comprehensive Analysis. FA Davis Company; 1983

[35] de Visser E, Pauwels J, Duysens JEJ, Mulder T, Veth RPH. Gait adaptations during walking under visual and cognitive constraints: A study of patients recovering from limb-saving surgery of the lower limb1. American Journal of Physical Medicine & Rehabilitation. 1998;**77**(6):503-509

[36] Maria M. Martins, Anselmo Frizera Neto, Cristina Santos, et Ramón Ceres. Review and classification of human gait training and rehabilitation devices. Assistive Technology Research Series, 29: 774–781, 2011.

[37] Duysens J, Van de Crommert HWAA. Neural control of locomotion; Part 1: The central pattern generator from cats to humans. Gait & Posture. 1998;**7**(2):131-141

[38] Defebvre L. Troubles de la marche. EMC-Traité Médecine AKOS. 2010;**5**:1-7

[39] Rasch PJ, Burke RK. Kinesiology and Applied Anatomy: The Science of Human Movement. Lea & Febiger; 1978

[40] Inman VT, Ralston HJ, Todd F. Human Walking. Williams & Wilkins; 1981

[41] Malfait B, Dingenen B, Smeets A, Staes F, Pataky T, Robinson MA, et al. Knee and hip joint kinematics predict quadriceps and hamstrings neuromuscular activation patterns. PLoS one. 2016;**11**(4):e0153737

[42] Shultz SJ, Houglum PA, Perrin DH. Examination of musculoskeletal injuries. Human Kinetics. 2015

[43] Loudon JK, Swift M, Bell S. Human Kinetics. In: The Clinical Orthopedic Assessment Guide. 2008

[44] Langer S. A Practical Manual of Clinical Electrodynography. Langer Foundation for Biomechanics and Sports Medicine Research; 1989

[45] Terrier P, Schutz Y. How useful is satellite positioning system (GPS) to track gait parameters? A review. Journal of Neuroengineering and Rehabilitation. 2005;**2**(1):28

[46] Deckers J. Ganganalyse en loop training voor de paramedicus. Bohn Stafleu van Loghum; 1996

[47] Perry J, Davids JR. Gait analysis: normal and pathological function. Journal of Pediatric Orthopaedics. 1992; **12**(6):815

[48] Novacheck TF. The biomechanics of running. Gait & Posture. 1998;**7**(1):77-95

[49] Bowker JH, Hall CB. Normal human gait. In: Atlas of Orthotics. St. Louis: CV Mosby Co; 1975. pp. 133-143

[50] Gage JR. Gait Analysis in Cerebral Palsy. London: Mac Keith Press; 1991

[51] Berger W, Quintern J, Dietz V. Pathophysiology of gait in children with cerebral palsy. Electroencephalography and Clinical Neurophysiology. 1982;**53** (5):538-548

[52] Gatev V, Ivanov I. Excitation-contraction latency in human muscles. Agressologie. 1972;**13**(6):7-12

[53] Frost G, Dowling J, Dyson K, Bar-Or O. Cocontraction in three age groups of children during treadmill locomotion. Journal of Electromyography and Kinesiology. 1997;**7**(3):179-186

[54] Lamontagne A, Richards CL, Malouin F. Coactivation during gait as an adaptive behavior after stroke. Journal of Electromyography and Kinesiology. 2000;**10**(6):407-415

[55] Lametti DR, Houle G, Ostry DJ. Control of movement variability and the regulation of limb impedance. Journal of Neurophysiology. 2007;**98**(6):3516-3524

[56] Milner TE, Cloutier C, Leger AB, Franklin DW. Inability to activate muscles maximally during cocontraction and the effect on joint stiffness. Experimental Brain Research. 1995;**107**(2):293-305

[57] Milner TE. Adaptation to destabilizing dynamics by means of muscle cocontraction. Experimental Brain Research. 2002;**143**(4):406-416

[58] Zakotnik J, Matheson T, Dürr V. Co-contraction and passive forces facilitate load compensation of aimed limb movements. Journal of Neuroscience. 2006;**26**(19):4995-5007

[59] Higginson JS, Zajac FE, Neptune RR, Kautz SA, Delp SL. Muscle contributions to support during gait in an individual with post-stroke hemiparesis. Journal of biomechanics. 2006;**39**(10):1769-1777

[60] Darainy M, Ostry DJ. Muscle cocontraction following dynamics learning. Experimental Brain Research. 2008;**190**(2):153-163

[61] Feldman AG, Levin MF. The origin and use of positional frames of reference in motor control. Behavioral and Brain Sciences. 1995;**18**(4):723-744

[62] Bhushan N, Shadmehr R. Computational nature of human adaptive control during learning of reaching movements in force fields. Biological Cybernetics. 1999;**81**(1):39-60

[63] Gribble PL, Ostry DJ. Compensation for interaction torques during single and multi-joint limb movement. Journal of neurophysiology. 1999;**82**(5):2310-2326

[64] Neilson PD, Neilson MD. An overview of adaptive model theory: Solving the problems of redundancy, resources, and nonlinear interactions in human movement control. Journal of Neural Engineering. 2005;**2**(3):S279

[65] Todorov E. Direct cortical control of muscle activation in voluntary arm movements: A model. Nature neuroscience. 2000;**3**(4):391

[66] Busse ME, Wiles CM, Van Deursen RWM. Muscle co-activation in neurological conditions. Physical Therapy Reviews. 2005;**10**(4):247-253

[67] Olney SJ. Quantitative evaluation of co-contraction of knee and ankle muscles in normal walking. In: Biomechanics IX-A. Champain, IL: Human Kinetics; 1985. pp. 431-437

[68] Smith AM. The coactivation of antagonist muscles. Canadian Journal of Physiology and Pharmacology. 1981;**59**(7):733-747

[69] Damiano DL. Reviewing muscle cocontraction: Is it a developmental, pathological, or motor control issue? Physical & Occupational Therapy in Pediatrics. 1993;**12**(4):3-20

[70] Barnet CH, Harding D. The activity of antagonist muscles during voluntary movement. Rheumatology. 1955;**2**(8):290-293

[71] Cheng C-H, Lin K-H, Wang J-L. Co-contraction of cervical muscles during sagittal and coronal neck motions at different movement speeds. European Journal of Applied Physiology. 2008;**103**(6):647

[72] Lestienne F, Bouisset S. Temporal pattern of the activation of an agonist and antagonist as a function of the tension of the agonist. Revue Neurologique. 1968;**118**(6):550-554

[73] Patton NJ, Mortensen OA. An electromyographic study of reciprocal activity of muscles. The Anatomical Record. 1971;**170**(3):255-268

[74] Kellis E, Unnithan VB. Co-activation of vastus lateralis and biceps femoris muscles in pubertal children and adults. European Journal of Applied Physiology and Occupational Physiology. 1999; **79**(6):504-511

[75] Baratta R, Solomonow M, Zhou BH, Letson D, Chuinard R, D'ambrosia R. Muscular coactivation: The role of the antagonist musculature in maintaining knee stability. The American Journal of Sports Medicine. 1988;**16**(2):113-122

[76] Osternig LR, Caster BL, James CR. Contralateral hamstring (biceps femoris) coactivation patterns and anterior cruciate ligament dysfunction. Medicine and Science in Sports and Exercise. 1995; **27**(6):805-808

[77] Unnithan VB, Dowling JJ, Frost G, Ayub Volpe B, Bar-Or O. Cocontraction and phasic activity during GAIT in children with cerebral palsy. Electromyography and clinical neurophysiology. 1996;**36**(8):487-494

[78] Peterson DS, Martin PE. Effects of age and walking speed on co-activation and cost of walking in healthy adults. Gait & Posture. 2010;**31**(3):355-359

[79] Spiegel KM, Stratton J, Glendinning DS, Enoka RM. The influence of age on the assessment of motor unit activation in a human hand muscle. Experimental Physiology: Translation and Integration. 1996;**81**(5):805-819

[80] Fonseca ST, Silva PLP, Ocarino JM, Ursine PGS. Analysis of an EMG method for quantification of muscular co-contraction. Rev Bras Ciên e Mov. 2001; **9**(3):23-30

[81] Hesse S, Brandl-Hesse B, Seidel U, Doll B, Gregoric M. Lower limb muscle activity in ambulatory children with cerebral palsy before and after the treatment with Botulinum toxin A. Restorative Neurology and Neuroscience. 2000;**17**(1):1-8

[82] Concato J. Observation alversus experimental studies: What's the evidence for a hierarchy? NeuroRx. 2004;**1**(3):341-347

[83] Massaad F, Lejeune TM, Detrembleur C. Reducing the energy cost of hemiparetic gait using center of mass feedback: A pilot study. Neurorehabilitation and Neural Repair. 2010;**24**(4):338-347

[84] Lenzi T, De Rossi SMM, Vitiello N, Carrozza MC. Intention-based EMG control for powered exoskeletons. IEEE Transactions on Biomedical Engineering. 2012;**59**(8):2180-2190

[85] Falconer K, Winter DA. Quantitative assessment of cocontraction at the ankle joint in walking. Electromyography and Clinical Neurophysiology. 1985;**25**(2-3): 135-149

[86] Den Otter AR, Geurts ACH, Mulder T, Duysens J. Speed related changes in muscle activity from normal to very slow walking speeds. Gait & Posture. 2004; **19**(3):270-278

[87] Choukri K, Chollet G. Adaptation of automatic speech recognizers to new speakers using canonical correlation analysis techniques. Computer Speech & Language. 1986;**1**(2):95-107

[88] Knarr BA, Zeni Jr JA, Higginson JS. Comparison of electromyography and joint moment as indicators of co-contraction. Journal of Electromyography and Kinesiology. 2012;**22**(4):607-611

[89] Knutsson E, Richards C. Different types of disturbed motor control in gait of hemiparetic patients. Brain: A Journal of Neurology. 1979;**102**(2):405-430

[90] Detrembleur C, Dierick F, Stoquart G, Chantraine F, Lejeune T. Energy cost, mechanical work, and efficiency of hemiparetic walking. Gait & Posture. 2003;**18**(2):47-55

[91] Den Otter AR, Geurts ACH, Mulder TH, Duysens J. Gait recovery is not associated with changes in the temporal patterning of muscle activity during treadmill walking in patients with post stroke hemiparesis. Clinical Neurophysiology. 2006;**117**(1):4-15

[92] Den Otter AR, Geurts ACH, Mulder TH, Duysens J. Abnormalities in the temporal patterning of lower extremity muscle activity in hemiparetic gait. Gait & Posture. 2007;**25**(3):342-352

[93] Chow JW, Yablon SA, Stokic DS. Coactivation of ankle muscles during stance phase of gait in patients with lower limb hypertonia after acquired brain injury. Clinical Neurophysiology. 2012;**123**(8):1599-1605

[94] Derouesne C, Cambon H, Yelnik A, Duyckaerts C, Hauw JJ. Infarcts in the middle cerebral artery territory: Pathological study of the mechanisms of death. Acta neurologica scandinavica. 1993;**87**(5):361-366

[95] Damiano DL, Martellotta TL, Sullivan DJ, Granata KP, Abel MF. Muscle force production and functional performance in spastic cerebral palsy: Relationship of cocontraction. Archives of Physical Medicine and Rehabilitation. 2000;**81**(7):895-900

[96] Charles T. Leonard, Helga Hirschfeld, and Hans Forssberg. The Development of Independent Walking in Children With Cerebral Palsy. Developmental Medicine & Child Neurology, 33(7):567–577, 1991

[97] Keefer DJ, Wayland Tseh JL, Caputo K, Apperson S, McGreal PV, Morgan DW. Interrelationships among thigh muscle co-contraction, quadriceps muscle strength and the aerobic demand of walking in children with cerebral palsy. Electromyography and Clinical Neurophysiology. 2004;**44**(2):103-110

[98] Prosser LA, Lee SCK, VanSant AF, Barbe MF, Lauer RT. Trunk and hip muscle activation patterns are different during walking in young children with and without cerebral palsy. Physical Therapy. 2010;**90**(7):986-997

[99] Wakeling J, Delaney R, Dudkiewicz I. A method for quantifying dynamic muscle dysfunction in children and young adults with cerebral palsy. Gait & Posture. 2007;**25**(4):580-589

[100] Fonseca ST, Silva PLP, Ocarino JM, Guimaraes RB, Oliveira MTC, Lage CA. Analyses of dynamic cocontraction level in individuals with anterior cruciate ligament injury. Journal of Electromyography and Kinesiology. 2004;**14**(2):239-247

[101] Merletti R, Di Torino P. Standards for reporting EMG data. J Electromyogr Kinesiol. 1999;**9**(1):3-4

[102] Burden AM, Trew M, Baltzopoulos V. Normalisation of gait EMGs: A reexamination. Journal of Electromyography and Kinesiology. 2003;**13**(6):519-532

[103] Shiavi R, Frigo C, Pedotti A. Electromyographic signals during gait: Criteria for envelope filtering and number of strides. Medical and Biological Engineering and Computing. 1998;**36**(2):171-178

[104] Mohamad K, Nahla T, Samer ALFAYAD, Ouezdou FB, Chitour Y, Dychus E. Mechanical development of a scalable structure for adolescent exoskeletons. In: 2019 IEEE 16th International Conference on Rehabilitation Robotics (ICORR). IEEE; 2019. pp. 323-330

Chapter 3

Design and Control of the McKibben Artificial Muscles Actuated Humanoid Manipulator

Daoxiong Gong and Jianjun Yu

Abstract

The McKibben Pneumatic Artificial Muscles (PAMs) are expected to endow the advanced robots with the ability of coexisting and cooperating with humans. However, the application of PAMs is still severely hindered by some critical issues. Focusing on the bionic design issue, this chapter in detail presents the design of a 7-degree-of-freedom (DOF) human-arm-like manipulator. It takes the antagonized PAMs and Bowden cables to mimic the muscle-tendon-ligament structure of human arm by elaborately configuring the DOFs and flexibly deploying the routing of Bowden cables; as a result, the DOFs of the analog shoulder, elbow, and wrist of the robotic arm intersect at a point respectively and the motion of these DOFs is independent from each other for convenience of human-like motion. The model imprecision caused by the strong nonlinearity is universally acknowledged as a main drawback of the PAM systems. Focusing on this issue, this chapter views the model imprecision as an internal disturbance, and presents an approach that observe these disturbances with extended-state-observer (ESO) and compensate them with full-order-sliding-mode-controller (fSMC), *via* experiments validated the human-like motion performance with expected robustness and tracking accuracy. Finally, some variants of PAMs for remedying the drawbacks of the PAM systems are discussed.

Keywords: McKibben artificial muscles, bionic design, anthropomorphic arm, human-like motion, full-order sliding mode control, extended state observer

1. Introduction

Advanced robots, especially the humanoids, are expected to be employed as cooperative robot for coexisting and collaborating with human and interacting with other robots [1] in the human-centered, complex, unstructured, and dynamic daily-living environment, using the tools and equipment designed for human to accomplish some complex tasks or to be adopted as the slave robot to facilitate the tele-manipulation because its human-like configurations is favorable for simplifying the motion mapping as well as improving the efficiency of a homogeneous master-slave system.

Such an ideal humanoid robot must have muscle-like actuators, bionic musculoskeletal structure (appearance), and human-like motion patterns (behaviors). This chapter will focus and highlight some insights on these topics.

1.1 The McKibben artificial muscles as the bionic actuator

The McKibben Pneumatic Artificial Muscle (PAM) is widely regarded as a promising actuator for robotics and automation [2], and extensive research studies have been conducted to improve its applications in humanoid robot systems. A PAM consists mainly of a rubber bladder with reinforcement knitting nested outside or embedded in it (**Figure 1**). When actuated with a supply of compressed air, it extends in the circumferential direction and contracts in length; as a result, it is able to generate a tensile force high up to hundreds or even thousands Newton and a contraction motion in the longitudinal direction up to 25–37% of its original length. The PAM generates both powerful tensile force and good responsive movement at the start of the contraction and then decreases as it contracts, which behavior is similar to those of a biological muscle.

In a typical situation, a PAM requires a pressure air source and two pneumatic valves, one of which as the inlet valve for filling the air into the PAM, and the other as the outlet valve for exhausting the air out. By controlling the opening of both valves, it is possible to realize the desired contraction/force of the PAM for robot actuation [3].

Because a PAM can only generate a contraction force, people usually adopt a pair of PAMs to mimic the antagonistic configuration (**Figure 2b**) of a human muscle pair actuating a human joint, e.g., the antagonistic muscles pair of the biceps and the triceps brachii efficiently realize the bidirectional motion of the elbow joint, i.e., the

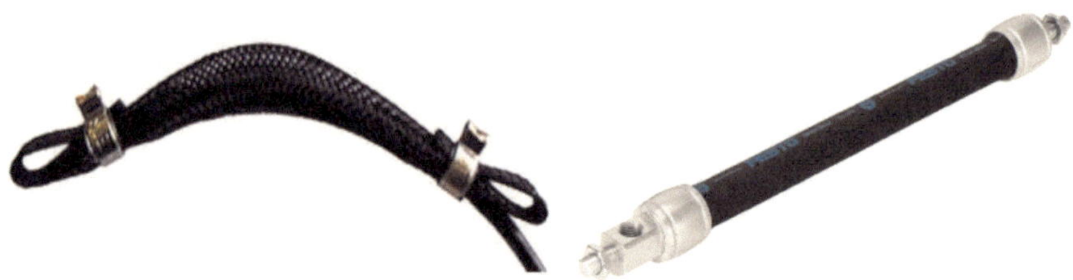

Figure 1.
Two kinds of the Mckibben artificial muscles (with reinforcement nested outside or embedded in).

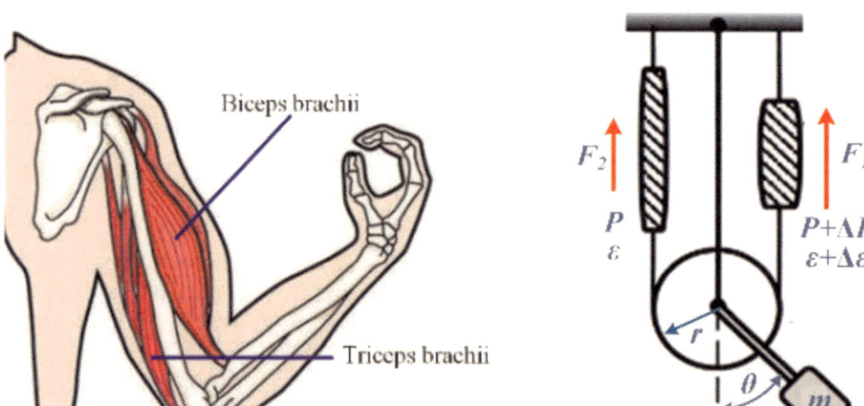

Figure 2.
The antagonistic muscles-driven strategy. (a) The human joint; (b) the robotic joint acutated by PAMs via steel wire cables.

elbow flexes when the biceps contracts and the triceps relax, and the elbow straightens when the biceps relaxes and the triceps contracts (**Figure 2a**). By combining the PAMs and the steel wire rope as an artificial muscle-tendon structure and attaching the ropes on the pulley circumference of the robotic joint, people can conveniently keep the direction of contraction/relaxation of the PAMs unchanged, avoid the misalignment of the PAMs, and keep the linear relationship between the contraction/relaxation length of PAMs to the rotation angle of the robotic joint:

$$\varepsilon = r\vartheta \tag{1}$$

where ε is the length of contraction/relaxation of the antagonistic PAM pair, r the pulley radius and ϑ the rotation angle of the pulley in radian, and it should cover the rotary scope of the concerned joint.

PAMs possess compelling properties in various aspects [4], including the inherent compliance, the actuating characteristics similar to those of the human muscle, high power-to-mass and power-to-volume ratio, flexibility of deployment, low cost, etc. Therefore, researchers have conducted vast studies in recent years and have gotten a lot of progress in the robot application study of PAMs.

1.2 The bionic design of humanoid robots actuated by PAMs

PAMs are widely regarded as an ideal actuator for cooperative robots because its actuating properties are similar to those of the human muscles and its inherent compliance stems from the mediator of compressible air. Therefore, people are prone to realize the human-like dexterous and compliant manipulation by mimicking both the organization and the actuating characteristics of the human muscles with PAMs.

Researchers studied the structure of human arm/hand and mimicking the configuration features in the robot designation to realize the human-like dexterity and have invented many anthropomorphic musculoskeletal robot systems. The famous FESTO Airic's arm [5] and SHADOW Hand [6] are two examples of the bionic musculoskeletal robot arm/hand, respectively. The FESTO Airic's arm imitates the structure of human arm with highly similarity. It consists of the ulna, the radius, the metacarpal and phalanges, a shoulder ball joint, and the scapular structure actuated by 30 FESTO Fluidic muscles to imitate the arrangement of the muscles of the human arm

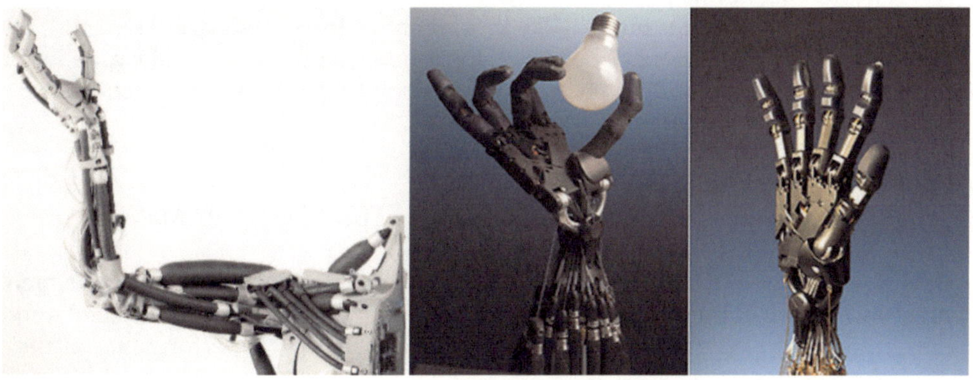

Figure 3.
The bionic musculoskeletal robots of the FESTO Airic's arm and the SHADOW hand [5, 6].

(**Figure 3** left). The Shadow Hand is designed to replicate the functionality, dimensions, and range of motion of the human hand (**Figure 3** right). With 20 DOFs actuated by PAMs as well as touch sensors mounted on the fingertips, the Hand as a slave robot end-effector is able to precisely mimic the master operator's movements to complete complex teleoperation tasks.

In order to achieve highly human-like dexterity, researchers even try to replicate the manipulation dexterity of human hand in the robotic system via highly biomimetic design. For example, Nanayakkara et al. reviewed the unique musculoskeletal structure of the human hand for inspiring the design of robotic hand to improve its dexterous grasping and manipulating capabilities [7]; Xu et al. designed a biomimetic robotic hand that closely mimics the human hand with artificial joint capsules, crocheted ligaments and tendons, laser-cut extensor hood, and elastic pulley mechanisms [8]; Faudzi et al. designed an index finger of human-like robotics hand that closely replicates the human finger in terms of bones, ligaments, thin McKibben PAMs, extensor mechanism, tendon, and pulley system [9]; Tasi et al. designed an anatomically correct, biomechatronic hand that emphasizes the anatomical consistency with the human hand and implements every functionally significant bone, ligament, finger-actuating intrinsic and extrinsic muscles, tendinous network, and pulley system to realize the same synergistic movements of the human hand [10]. Ikemoto et al. designed a musculoskeletal robot arm driven by PAMs to mimic the structure of the human shoulder complex and try to assure a wide range of movements and the joint's stability at the same time [11].

However, a robotic arm with highly similarity in construction to the human arm will certainly lead to a very complex and elaborated mechanical structure, which may pose a challenge to both the manufacture and the motion control of the system. Therefore, people should emphasize more similarity to human arm in functionality than in construction when they conceive the design scheme of the robotic arm, especially they should pay attention to the scheme that is able to eliminate the coupling effect among DOFs to facilitate the human-like motion control of the robotic arm.

Some of the human arm joints can be viewed as a spherical joint. A spherical joint actuated by muscle groups may lead to strong coupling between the tensile forces of the muscles, which will make the motion control of the joint difficult. People usually decompose the motion of a spherical joint into two or three orthogonal rotation axes with every rotation axis actuated by a pair of antagonistic muscles. However, if the robotic orthogonal rotation axes of a counterpart of the human joint (e.g., shoulder) are not intersecting at one point, the motion of different DOFs may become coupling and lead to unnatural or not human-like movement of the robotic arm [12, 13].

This is a critical issue need to be solved for promoting the PAM application in the humanoid robotic systems. This chapter will illustrate a solution to this issue via an example of robotic manipulator system.

1.3 The strong nonlinearity and modeling/control problem of PAMs

It is generally agreed that it is very difficult to model PAM precisely, and the model imprecision, which caused by both nonlinearities (including hysteresis effects) and the uncertainty of parameters, will severely impede the control performance of the PAM system.

Extensive studies have been conducted by researchers to improve the control performance of PAM systems.

As a widely regarded powerful robust nonlinear control scheme, the Sliding Mode Control (SMC) and its many varieties, e.g., the SMC [14], the second-order SMC [15], the hybrid of SMC and adaptive fuzzy Cerebellar Model Articulation Controller [16], the Nonsingular Terminal SMC [17], the fuzzy terminal SMC [18], etc., have been adopted to control the PAM systems and have shown a vast potential for the application of SMC in the PAM system. Besides the miscellaneous modeling and control methods, a lot of intelligent modeling methods and control algorithms, e.g., the modeling method based on machine learning [19] or artificial neural networks [20], as well as the control algorithms based on learning vector quantization neural network [21], fuzzy control [22], and nonparametric control algorithms [23], have been studied.

Though these studies have demonstrated some exciting results, the model inaccuracy problem of PAM device and the nonlinearity-caused control difficulty of the PAM system still need further studies. This chapter will present an effective solution to this problem via a case study.

1.4 The human-like motion control of the PAM systems

Human-like motions are critical for a robot to coexist and collaborate with human, as well as interact with other robots [1], in the human-centered, complex, unstructured, and dynamic daily-living environment. The anthropomorphic features of the robot and the human trust motion pattern of the robot have been experimentally validated to be very important for performing human-robot collaboration tasks [24]. A human cooperator will feel comfortable and friendly if the robot exhibits the human-like behaviors, which are predictable and understandable for the human partner involved in the collaboration [25]. Otherwise, an unexpected speed variation of a robotic arm will increase the "surprise" subjective rate, cause the mental strain of human cooperators, and reduce the efficiency of human-robot cooperation [26]. Therefore, it is important to make the robot's motion patterns natural, understandable, and predictable to the human cooperators for effective human-robot collaboration [27].

Studies in physiology, anatomy, biomechanics, and neuroscience have revealed that human motion has its own special characteristics [28]. The motion of human arm in the free space is assured satisfying the minimum jerk model, i.e., the motion of the human arm conforms to the Logistic Equation and shows a bell-shaped velocity profile [29].

To realize the expected human-like motion on a robotic manipulator actuated by PAMs is a tracking control problem of the nonlinear robotic system. This chapter will concern and present a feasible solution to it.

2. Bionic design of humanoid manipulator

As mentioned above, a highly biomimetic design may lead to a very elaborated mechanical structure and pose a challenge to manufacture and motion control of the system. Especially, a spherical joint actuated by muscle groups may cause strong coupling between the tensile forces of the muscles and make the motion control difficult. To decompose the motion of a spherical joint into two or three orthogonal rotation axes is an option, but if these robotic orthogonal rotation axes of a spherical joint cannot intersect at one point, the motion of the robotic arm may become unnatural or not human-like and consequently degrade the robot system.

A design scheme of a human-arm-like manipulator actuated by antagonized PAM pairs that nicely solved this problem is presented in the following sections.

The following measures are taken to endow the humanoid robotic manipulator the ability of reproducing the motion of human arm/hand and easy cooperating with human:

1. The DOF configurations of the robotic arm are similar to those of human arm;

2. The mechanical structure should be as simple as possible for easy fabrication, assembling, and maintenance;

3. The motion of every DOF is independent from each other for reducing the effect of coupling and easy control.

2.1 The 7-DOF human-arm-like manipulator actuated by antagonized PAMs

2.1.1 The kinematic model of the robotic arm

The designed robotic arm possesses a shoulder joint with three DOFs (flexion/extension, abduction/adduction and internal/external rotation of the upper arm), and an elbow joint with two DOFs (flexion/extension and pronation/supination of the forearm) and a wrist joint with two DOFs (flexion/extension, abduction/adduction of the hand). The same DOF configuration, the same joint motion range (**Table 1**), as well as the same proportion of limb length of the Master Motor Map (MMM) kinematic model [30] are taken in the robot arm's design scheme [31]. As shown in **Figure 4**, 7 joints of the robotic arm are independently actuated by a pair of antagonistic PAMs (the circles and arrows in different color indicate the joints actuated by different PAMs as illustrated in **Figures 6–8**). This DOF configuration of the robotic arm makes sure that the three joints of its analog shoulder, the two joints of its analog elbow, as well as the two joints of its analog wrist intersect at a common point, respectively. These structural features make the robotic arm kinematically homogeneous to the human arm; as a result, the robot arm is able to reproduce the motion of human arm with the human-like flexibility, e.g., to recover from errors, to avoid singularity, and to perform obstacle avoidance. Furthermore, this DOF configuration as well as the mechanical structure of the joints of the robotic arm as descripted below will make the motion of the adjacent joints decoupled and consequently convenient for the motion control of the robotic arm.

2.1.2 The configuration of PAMs and cables in the robotic arm

Generally, an anthropomorphic robotic arm means that a very complex and elaborated mechanical structure is needed, and the whole system may become very difficult

Joint	DOF	X-limits	Z-limits	Y-limits
Shoulder	3	[−120°,90°]	[−90°,30°]	[−60°,60°]
Elbow	2	[0°,135°]	—	[−90°,90°]
Wrist	2	[−30°,30°]	[−60°,60°]	—

Table 1.
The range of motion of each DOF.

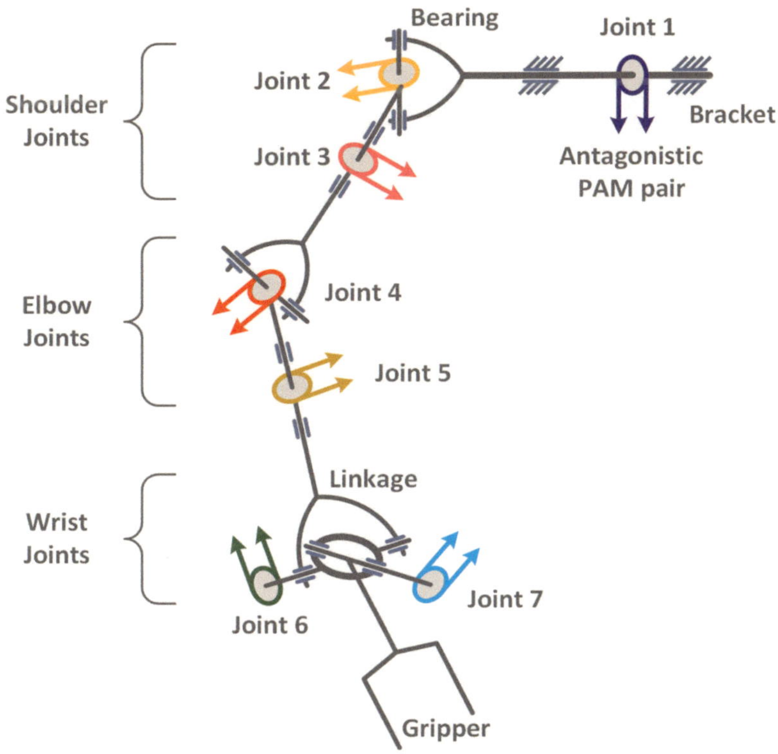

Figure 4.
The DOF configuration and the joint actuating scheme of the humanoid arm.

to control. In order to facilitate the actuation/control of the humanoid arm, Bowden cables are adopted to decouple the tensile forces of the different antagonistic PAM pairs.

The Bowden cable consists of an outer sheath and an inner tendon, the inner tendon (with diameter of 1.5 mm) is used for actuation transmission of every DOF, and the outer sheath is used for guiding the motion of the inner tendon in a curved path that gets over the moving adjacent DOF. As a result, the motion of the concerned DOF will not be disturbed by the motion of the other DOFs. The Bowden cable scheme has many advantages, e.g., lightweight, space saving, and especially the flexible layout of PAMs on the robot without suffering from misalignments, which make it a suitable solution for the PAMs actuation transmission of the robotic system.

The designed humanoid robotic arm is composed of a torso, an upper arm (386 mm in length), a forearm (342 mm in length), and a hand as shown in **Figure 5**. The pressure air source is placed under the table, and the valves and air pipelines are placed on the shelf behind the robot. Both of the upper arm and the forearm have a bar placed on the center to serve as the skeletons, and four PAMs are installed evenly around the bar as the muscles. The torso is an aluminum frame used to support the robotic arm, and six PAMs are installed on the torso as the muscles to actuate the shoulder joint. Bowden cables are adopted as the tendons to transmit the actuation force to the DOFs of the joint of the shoulder, the elbow, and the wrist, respectively.

Though the shoulder complex usually is considered as the most complicated part of the human body as well as with the greatest range of motion than any other joint in the body, the motion of the three DOFs of the robotic shoulder joint can be decoupled via the elaborative arrangement of Bowden wires.

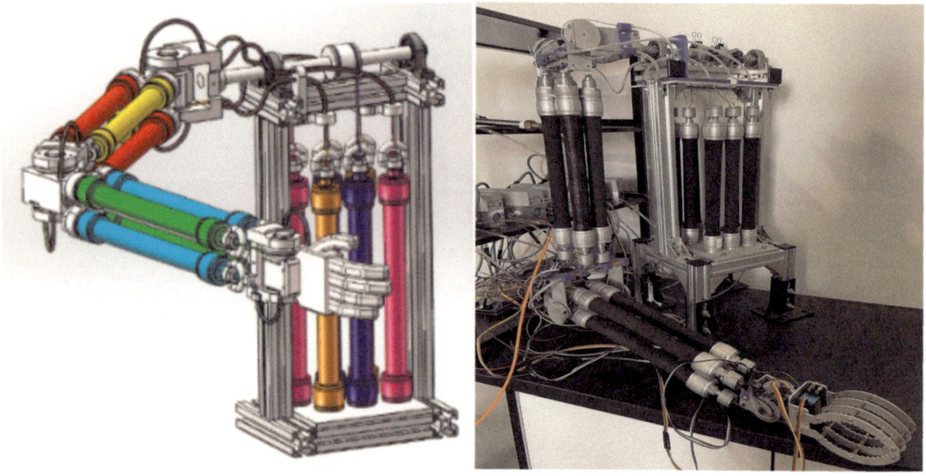

Figure 5.
The structure and the physical device of the 7-DOF humanoid arm actuated by PAMs [31].

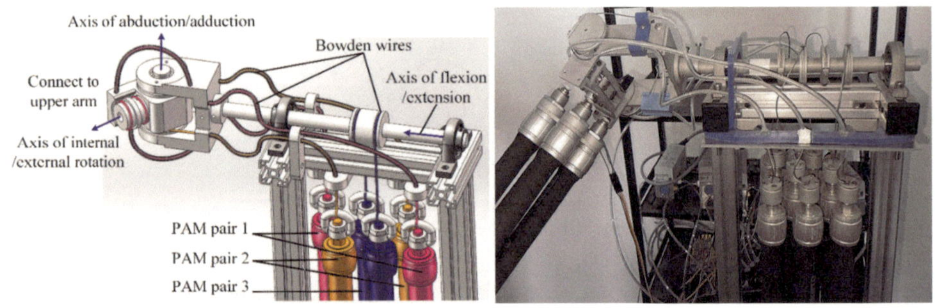

Figure 6.
The structure and the physical device of the shoulder joint [31].

As shown in **Figure 6**, the robotic shoulder consists of three DOFs, the first one for the flexion/extension of the upper arm (actuated by PAMs in blue), and the second one for the abduction/adduction of the upper arm (actuated by PAMs in orange), and the third one for the internal/external rotation of the upper arm (actuated by PAMs in magenta). One end of these PAMs is fixed on the framework of the torso and another end is attached to the corresponding axis of DOF via Bowden cables. The outer sheath in orange bypasses the DOF of flexion/extension and attached on the fixed part of the DOF of adduction/abduction, while the inner tendon is attached on the rotation part of the DOF of adduction/abduction. Also, the outer sheath in magenta bypasses the DOFs of both flexion/extension and adduction/abduction and attached on the fixed part of the DOF of internal/external rotation, while the inner tendon is attached on the rotation part of the DOF of internal/external rotation. Consequently, the motions of the three DOFs of the shoulder joint are independent from each other, and the control algorithm of these DOFs can be separately designed without needing to consider their coupling (or interaction) effect.

As shown in **Figure 7**, the elbow joint consists of two DOFs, one for the flexion/extension of the forearm (actuated by PAMs in red), and another for the pronation/supination of the forearm (actuated by PAMs in yellow). One end of these PAMs is

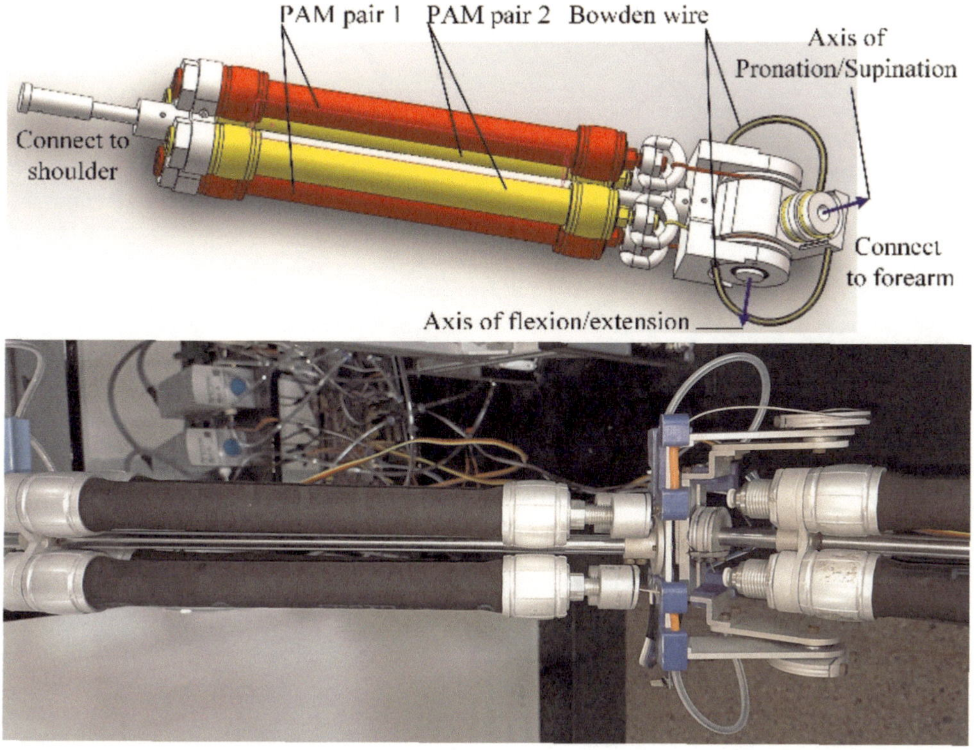

Figure 7.
The structure and the physical device of the elbow joint [31].

fixed on the framework of the upper arm and another end is attached to the corresponding axis of DOF via Bowden cables. The outer sheath in yellow bypasses the DOF of flexion/extension and attached on the fixed part of the DOF of pronation/supination, while the inner tendon is attached on the rotation part of the DOF of pronation/supination. Consequently, the motions of flexion/extension and pronation/supination of the elbow joint are independent from each other, and the control algorithm for these two DOFs can be separately designed without needing to consider their coupling (or interaction) effect.

As shown in **Figure 8**, the wrist joint consists of two DOFs, one for the flexion/extension of the hand (actuated by PAMs in green) and another for the abduction/adduction of the hand (actuated by PAMs in blue). One end of these PAMs is fixed on the framework of the forearm and another end is attached to the corresponding axis of DOF via Bowden cables. The outer sheath in blue bypasses the DOF of flexion/extension and attached on the fixed part of the DOF of abduction/adduction, while the inner tendon is attached on the rotation part of the DOF of abduction/adduction. Consequently, the motions of flexion/extension and abduction/adduction are independent from each other, and the control algorithm for these two DOFs can be separately designed without needing to consider their coupling (or interaction) effect.

Because the kinematic configuration of the designed robotic arm is the same as that of the MMM human arm reference model, the robotic arm is able to reproduce the motion of human arm. Due to the uniqueness of the joint design and Bowden cable arrangement, each DOF of the designed humanoid robotic arm is independently actuated by a pair of antagonistic muscles, and the motion of each DOF is decoupled

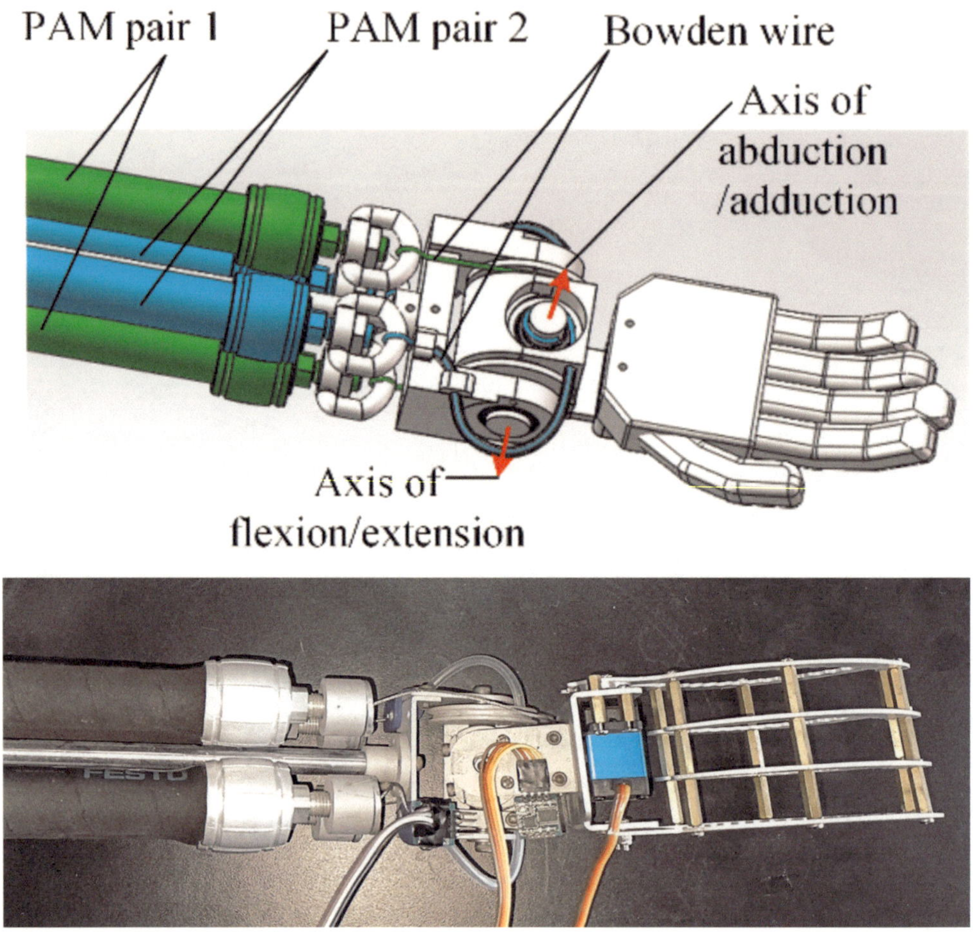

Figure 8.
The structure and the physical device of the wrist joint [31].

from each other. As a result, the mechanical structure of the robotic arm is simple, easy to be assembled, maintained, and controlled.

2.2 The design highlights for the anthropomorphic robotic hand

As shown in **Figure 8**, a gripper is attached on the robotic wrist. However, in order to endow the robotic systems the capability of human-like dexterous manipulation and easily cooperating with human, an anthropomorphic hand that mimics the functions of the human hand is essential, and the following factors should be taken into consideration in the design scheme of such a robotic hand [32]:

1. The robotic hand should be kinematically similar to the human hand (e.g., the MMM model) so that it is able to perform the same motions of the human hand. The DOF configuration of the human hand is very important for its delicate manipulations. For example, the abduction/adduction motion of the metacarpophalangeal (MCP) joint of the fingers (especially the thumb) for anthropomorphic grasping and the transverse metacarpal arch motion of the palms for scooping things. These DOF

configurations as well as their corresponding motion ranges should be carefully considered and realized in the robotic hand design.

2. The bionic design should focus on the similarity to human hand in functionality instead of construction, so that the mechanical structure is simple for easy fabrication, assembly, and maintenance and control. In order to do this, the aforementioned design scheme of the robotic arm can be used for reference of the design of the robotic hand, especially the DOF configurations as well as the routing of the Bowden cables for decoupling the motions of the orthogonal DOFs of the MCP joints. PAMs and Bowden cables are taken as analog of muscle-tendon-ligament structure to actuate each DOF of the robotic hand with a pair of antagonistic PAMs and Bowden cables for avoiding the motion coupling of adjacent joints. Though a human hand has both intrinsic and extrinsic muscles, all the PAMs should be arranged as extrinsic muscles and placed far away from the hand for not hindering the manipulation of the robotic hand because of the volume of the PAMs.

3. Air bladders can be attached at the robotic hand palm and finger segments as analog of the soft muscle pads and skin of human hand for mimicking the haptic sensing of the skins. Muscles and skin on the human hand bring it the capabilities of soft touch and tactile feedback, which is very important for dexterous robotic hand operating an object or interacting with human appropriately via motion/force control. Pneumatic tactile sensors based on air bladder possess some noticeable advantages for human-robot interaction application. The air bladder-based pneumatic tactile sensor is able to sense the lumping force acted on the chamber by measuring the pneumatic pressure change in the air bladder [33], and the air bladder itself is flexible for bending, stretching, and adapting to different mechanical structure on the robot. Regular textures can be designed on the bladders of the fingertip to mimic the fingerprint for enhancing the roughness sensing via the vibration signals generated when the fingertip slides on a rough surface. With the air bladder-based pneumatic tactile sensors as the pad of the fingers, the robotic hand is able to physically interact with humans and the environment via the "human-like compliant touch." Because of the inherent compliance of the mediator of compressible air in the bladder, the robotic hand is able to yield to unexpected impact when it collides with obstacles (environment or human) and then accordingly take measures to guarantee the safety of the robot system itself, the human, and the environment. With the air bladder-based pneumatic tactile sensors, the robotic hand may also explore the environment in a manner of "action for perception" [34] to obtain the information of the roughness of the object surface, the softness/stiffness of the object, and the slippage of an object grasped in the hand. Furthermore, the robotic hand is able to achieve a stable grasp as well as other manipulation tasks via force control based on the tactile sensing information.

2.3 Summary

Highly biomimetic design that mimics the musculoskeletal structure of human arm will inevitably lead to high structural complexity and control difficulty of the robotic system. Especially the spherical joint may lead to the motion of its decomposed orthogonal DOFs coupling, or lead to the movement of the robotic arm unnatural or not human-like because the rotation axes of these DOFs do not intersect at one point. Via the detailed bionic design of a 7 DOF human-arm-like manipulator, this section illustrated the DOFs' elaborate configuration and the Bowden cables' flexible routing scheme that successfully make the DOFs of a spherical joint intersected at a point and the motion of these DOFs independent from each other, so that the motion control and human-like dexterous manipulation can be conveniently realized on the PAMs actuated robotic system.

This example also exhibited the advantage of flexible layout of the PAMs in the actuation of robotic systems. If motors instead of PAMs are taken as the actuators of the robotic arm, it will be very difficult to make the DOFs of a spherical joint to intersect at a point, and the robotic arm will be offset and bulky because of the volume issue of the motors and reduction gears, which must be coaxially mounted on the axis of the DOF. While the PAMs can be placed far away from its actuated DOF, therefore the robotic manipulator actuated by PAMs may be small in size and light in weight for dexterous and convenient manipulations.

3. Human-like motion control for a PAMs-actuated robotic joint

As mentioned above, it is universally acknowledged that the difficulty of controlling a PAM system lies in the strong nonlinear characteristics of PAMs, and advanced control algorithms are urgently needed for making the best use of the advantages and bypassing the disadvantages of PAMs. In order to solve this problem, an fSMC-ESO algorithm [35] that combines the Extended State Observer (ESO) and the full-order Sliding Mode Controller (fSMC) for a robotic joint actuated by a pair of antagonistic PAMs is presented in the following sections. In this fSMC-ESO algorithm, the fSMC is adopted to eliminate the chatter and guaranteeing the finite-time convergence and the ESO to observe both the total disturbance and the states of the robot system. Experiments validate that the fSMC-ESO algorithm can efficiently inhibit the disturbance and compensate both the nonlinearity and the modeling inaccuracy to achieve good control performance of the PAM system.

3.1 The human-like motion pattern

According to the study of Flash et al. [36], assuming the movement to start and end with zero velocity and acceleration, the free motion trajectory of the human hand satisfies the minimum jerk model and conforms the Logistic Equation as follows:

$$\begin{cases} x(t) = x_0 + (x_0 - x_f)(-10\tau^3 + 15\tau^4 - 6\tau^5) \\ y(t) = y_0 + \left(y_0 - y_f\right)(-10\tau^3 + 15\tau^4 - 6\tau^5) \end{cases} \tag{2}$$

where $\tau = t/t_f$ is time variable, (x_0, y_0), $\left(x_f, y_f\right)$ are the position of the human hand at the initial time $(t = 0)$ and the final time $\left(t = t_f\right)$, and $-10,\ +15,\ -6$ are constant coefficients, respectively.

Consequently, for the motion of a robotic joint's, the desired value of the joint angle x_d, the angular velocity $\dot{x}_d$, and the angular acceleration $\ddot{x}_d$ will change over time according to the following human-like motion equations:

$$\begin{cases} x_d = x_f(10\tau^3 - 15\tau^4 + 6\tau^5) \\ \dot{x}_d = x_f(30\tau^2 - 60\tau^3 + 30\tau^4)/t_f \\ \ddot{x}_d = x_f(60\tau - 180\tau^2 + 120\tau^3)/t_f{}^2 \end{cases} \quad (3)$$

where x_f is the maximum rotation angle of the joint. The initial value of joint angle, the initial/final angular velocity, and the initial/final angular acceleration are zero.

3.2 Mathematical model of a robotic joint actuated by antagonistic PAMs

As illustrated in **Figures 4–8**, every joint of the designed robotic arm is independently actuated by a pair of antagonistic PAMs because the motions of adjacent joints are decoupled by the arrangement of Bowden cables. Therefore, this chapter only presents the human-like motion control scheme for a single joint system as shown in **Figure 9**. This experiment platform is a simplification of the robotic arm system of **Figure 5**. With only a single joint actuated by a pair of antagonistic PAMs, it simulates the weight lifting motion via the upward flexion of the elbow joint or the abduction of the shoulder with the elbow extended. The following control scheme for single joint can be easily promoted to the multiple joints of the robotic arm since the motions of its joints are independent from each other.

For the bionic robotic joint actuated by antagonistic PAMs of **Figure 9**, its mathematical model based on Yu's PAM model [37] is:

$$F(\varepsilon, p) = k_1(p) - k_2(p)\varepsilon + k_3(p)\exp(-\mu\varepsilon) \quad (4)$$

Figure 9.
The experimental platform of the bionic robotic joint actuated by antagonistic PAM pair.

where F is the output force of PAM; $\varepsilon = (L_0 - L)/L_0$ is the contraction rate of PAM; L_0, L are the original length and the actual length of PAM, respectively; p is the inner pressure of pneumatic muscles; $k_i(p) = k_{i1}p + k_{i2}, (i = 1, 2)$ are pending coefficients depended on p, k_{i1} and k_{i2} are fitting coefficients that are different in the inflation and the deflation processes; μ is the nonlinear attenuation coefficient of ε. Yu's PAM model consists of two parts, i.e., the model of the inflation process and those of the deflation process, and there is an obvious hysteresis between the inflation and deflation processes. In this study, only the inflation part of the Yu's model is adopted to modeling the robotic joint for convenient of computation, and the compensation of the model inaccuracy of the robotic joint (including those caused by the incompleteness of the PAM model) relies on the control algorithm.

Rewritten Eq. (4) as:

$$F(\varepsilon, p) = (k_{11}p + k_{12}) - (k_{21}p + k_{22})\varepsilon + (k_{31}p + k_{32})\exp(-\mu\varepsilon) = g(\varepsilon)p + h(\varepsilon) \quad (5)$$

where

$$\begin{aligned} g(\varepsilon) &= k_{11} - k_{21}\varepsilon + k_{31}\exp(-\mu\varepsilon) \\ h(\varepsilon) &= k_{12} - k_{22}\varepsilon + k_{32}\exp(-\mu\varepsilon) \end{aligned} \quad (6)$$

The dynamic equation of the bionic robotic joint is as follows:

$$J\ddot{\theta} + f_b\dot{\theta} = (F_1 - F_2)r - mgl\sin(\theta) + d \quad (7)$$

where $J = m_{pulley}r^2/2 + ml^2$ is the equivalent moment of inertia of the bionic joint, m_{pulley} is the mass of the pulley; m is the mass of the load; r is the radius of the pulley; l is the equivalent length of the connecting rod; θ is the rotation angle of the joint; $f_b\dot{\theta}$ is the damping term of the joint where the f_b is damping coefficient; F_1 and F_2 are the forces exerted by the antagonistic PAMs; d are the (external and internal) disturbances caused by the system parameter uncertainties.

Substituting the forces (5) of the antagonistic PAMs into (7):

$$J\ddot{\theta} + f_b\dot{\theta} = (g(\varepsilon_1)p_1 - g(\varepsilon_2)p_2 + h(\varepsilon_1) - h(\varepsilon_2))r - mgl\sin(\theta) + d \quad (8)$$

The antagonistic forces exerted by each PAM of the antagonistic pair should remain greater than 0 at all time, i.e., $F_1(t) > 0$, $F_2(t) > 0$, for $t \geq 0$. Therefore, the actual lengths of the PAMs are:

$$\begin{cases} L_1 = L_{1init} - r\theta \\ L_2 = L_{2init} - r\theta \end{cases} \quad (9)$$

where L_{1init}, L_{2init} are the initial lengths of the antagonistic PAMs when the rotation angle of the joint is $\theta = 0$.

Select the state variables as $x_1 = \theta$ and $x_2 = \dot{\theta}$, then the state equations of the studied joint are:

$$\begin{cases} \dot{x}_1 = x_2 \\ \dot{x}_2 = \dfrac{(h(\varepsilon_1) - h(\varepsilon_1))r - mgl\sin(x_1) - f_b x_2}{J} + \dfrac{(g(\varepsilon_1)p_1 - g(\varepsilon_2)p_2)r}{J} + \dfrac{d}{J} \end{cases} \quad (10)$$

Formulating $f(x) = [(h(\varepsilon_1) - h(\varepsilon_2))r - mgl\sin(x_1) - f_b x_2]/J$, $b = r/J$, $d_J = d/J$, and $u = g(\varepsilon_1)p_1 - g(\varepsilon_2)p_2$ as the control input, then,

$$\begin{cases} \dot{x}_1 = x_2 \\ \dot{x}_2 = f(x) + bu + d_J \end{cases} \tag{11}$$

This is a second-order nonlinear system with its control inputs u decided by the inner pressures of the antagonistic pair of PAMs, i.e., p_1 and p_2, which can be controlled to achieve the expected human-like motion of the studied robotic joint.

3.3 The full-order sliding mode control with extended state observer

The fSMC sliding surface is adopted as follows [38]:

$$s = \ddot{e} + c_2 sign(\dot{e})|\dot{e}|^{\alpha_2} + c_1 sign(e)|e|^{\alpha_1} \tag{12}$$

where constants $c_1 = 16$, $c_2 = 6$, $\alpha_1 = 5/8$, $\alpha_2 = 5/11$, and $\dot{e} = \dot{x}_d - \dot{x}_1$, $\ddot{e} = \ddot{x}_d - \ddot{x}_1$ are the first and second derivatives of e, $e = x_d - x_1$ is the tracking error of the joint, x_d, x_1 are the expected and the actual value of the joint angle, respectively.

The control input u is [38]:

$$u = -b^{-1}(u_{eq} + u_n) \tag{13}$$

where

$$u_{eq} = -\ddot{x}_d + f(x) - c_2 sign(\dot{e})|\dot{e}|^{\alpha_2} - c_1 sign(e)|e|^{\alpha_1} \tag{14}$$

$$\dot{u}_n + Tu_n = -\eta sign(s) \tag{15}$$

The Eq. (15) can serve as a low-pass filter to inhibit the high-frequency disturbance and to enhance the robustness of the system. Where $T = 0.01Hz$ is the bandwidth of the low-pass filter, and $\eta = 5$ is the control gain term.

Based on Eq. (11), an extended state $x_3 = f(x) + d_J$ is defined as the total disturbance (including both the external and the internal disturbances) estimated by the ESO, and consequently, a new model of the studied robotic joint system can be described with the new set of state variables:

$$\begin{cases} \dot{x}_1 = x_2 \\ \dot{x}_2 = x_3 + bu \\ \dot{x}_3 = \dot{f}(x) + \dot{d}_J \\ y = x_1 \end{cases} \tag{16}$$

Define z_1, z_2 and z_3 as the estimated value of the state variable x_1, x_2 and x_3, respectively, then the model of the tracking control system is:

$$\begin{cases} e_1 = z_1 - y \\ \dot{z}_1 = z_2 - \beta_{01}e_1 \\ \dot{z}_2 = z_3 - \beta_{02}e_1 + bu \\ \dot{z}_3 = -\beta_{03}e_1 \end{cases} \tag{17}$$

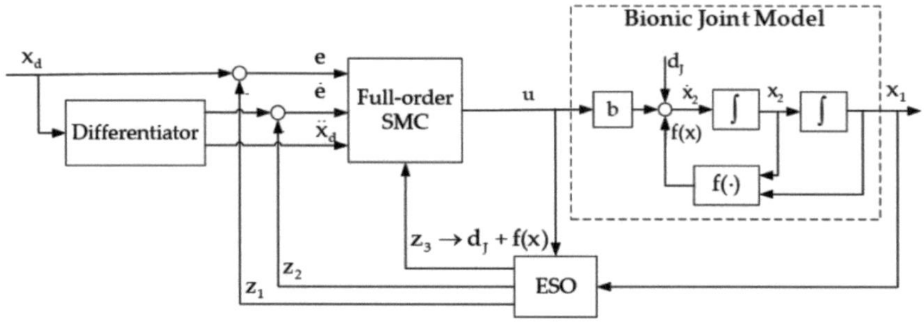

Figure 10.
Block diagram of the fSMC-ESO tracking control system.

where

$$[\beta_{01}\ \beta_{02}\ \beta_{03}]^T = \left[3\omega\ 3\omega^2\ \omega^3\right]^T \tag{18}$$

are positive constants and $\omega = 150\text{rad/s}$ according to reference [39].

As descripted above, the fSMC-ESO scheme views the model inaccuracy as a kind of internal disturbances and combines ESO and fSMC for achieving the human-like motion control of the robotic joint actuated by a pair of antagonistic PAMs.

The block diagram of the studied tracking control system is shown in **Figure 10**. In **Figure 10**, x_d and $\ddot{x}_d$ are the desired joint angle and the angular acceleration, which are computed with Eq. (3), of the human-like motion of the concerned robotic joint, respectively; e and $\dot{e}$ are the tracking error and its first derivative of the joint as explained in Eq. (12); the full-order SMC block consists of the Eqs. (12)–(15), which is chattering free and especially suiting for higher-order nonlinear systems with both uncertainty parameters and external disturbance [38]; the Bionic Joint Model block is the mathematic model of the concerned robotic joint as described in Eq. (16); and the ESO block is the extended state observer of the Active Disturbance Rejection Control [40], which is adopted to observe both the states of the robot system (z_1 and z_2) and the total disturbance z_3 of the robotic joint system.

Consequently, the fSMC-ESO control system for human-like motion of the joint can be descripted by Eqs. (17) and (18), and it is able to inhibit the disturbance and compensate the nonlinearity of the PAM system and achieve the expected tracking control performance with the control input:

$$u = -b^{-1}\left(-\ddot{x}_d + z_3 - c_2 sign(\dot{e})|\dot{e}|^{\alpha_2} - c_1 sign(e)|e|^{\alpha_1} + u_n\right) \tag{19}$$

3.4 Experimental validations

The proposed approach is validated via physical experiments on the platform of **Figure 9**. The PAMs are FESTO DMSP-20-200 with pre-tension of 40 N, and the parameters of the PAM model as well as those of the robotic joint are presented in **Tables 2** and **3**, respectively. In the physical experiments, the sampling time is

Parameter	k_{11}	k_{12}	k_{21}	k_{22}	k_{31}	k_{32}	L_0 (m)
Value	2.573	−222.7	6.389	1347.3	0.296	141.5	0.2

Table 2.
Parameters of mathematical model of PAMs.

Parameter	m_{pulley} (kg)	r (m)	m (kg)	l (m)	f_b $(N\ s/m)$
Value	0.2	0.016	3	0.2	0.2

Table 3.
Parameters of the robotic joint.

$0.01s$, $x_f = 90°$, $t_f = 2s$, and the force of the antagonistic muscle is constant $F_2 \equiv 40N$.

The performance of the fSMC-ESO approach against disturbance is validated via simulations of no external disturbance ($d = 0$), constant external disturbance ($d = 2rad$), and variable external disturbances (sinusoidal signal, $d = 2\sin(2\pi t)$, and $d = 2\sin(\pi t)$). The simulation results (both the output of the control system and the result of state observation) are presented in **Figure 11**, which validates that the ESO can accurately observe the state variation (including the external disturbances) of the system; as a result, the bionic joint is able to track the desired motion curve under different disturbances with an acceptable small error amplitudes under sinusoidal disturbances (0.006 and 0.001 of $d = 2\sin(2\pi t)$ and $d = 2\sin(\pi t)$, respectively) .

Then, the robustness of the proposed fSMC-ESO approach to the payload variation is validated in **Figure 12**, which illustrates the tension, the pressure, and the length of the antagonistic PAMs in the course of human-like motion with payload of 1 kg, 3 kg, and 5 kg, respectively. The fSMC-ESO approach can adjust the air pressure of PAMs according to load variation; as a result, it is robust to the payload variation and at the same time keeps good performance of human-like motion.

Figure 13 illustrates the physical experiment result of the robotic joint system, which achieves the expected human-like motion with a small overshoot (less than 1.2%, i.e., 1.1 degrees of angle). An estimated initial value of the total disturbance ($z_3(0) = 130$), which can be approximately determined via simulations and accordingly adjusted in the experiment, is applied to suppress the fluctuations at the initial period of the rise time so as to improve the performance of the control system. At the time of 6.5 s, an external disturbance is exerted on the robotic arm, and the control system is able to suppress the disturbance and drive the platform back to its primary position within 1 s. The good performances of trajectory tracking and robustness of the robotic joint system are reasonably attributed to both the state observing ability of ESO and the nonlinear control ability of fSMC.

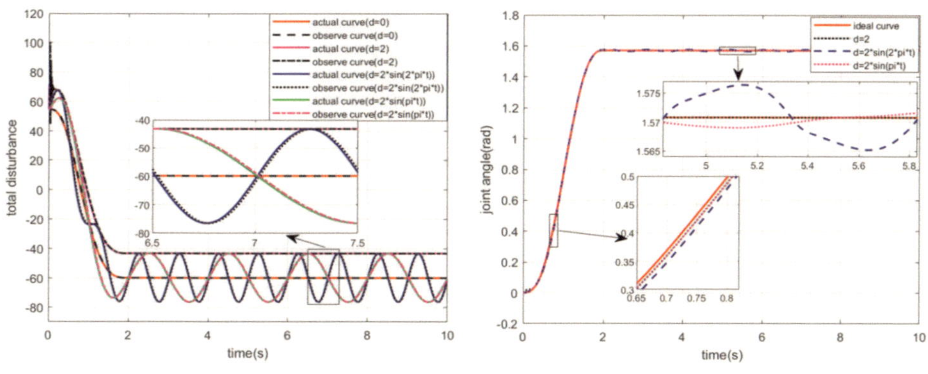

Figure 11.
Simulation of observation result and the joint angle response under different disturbances.

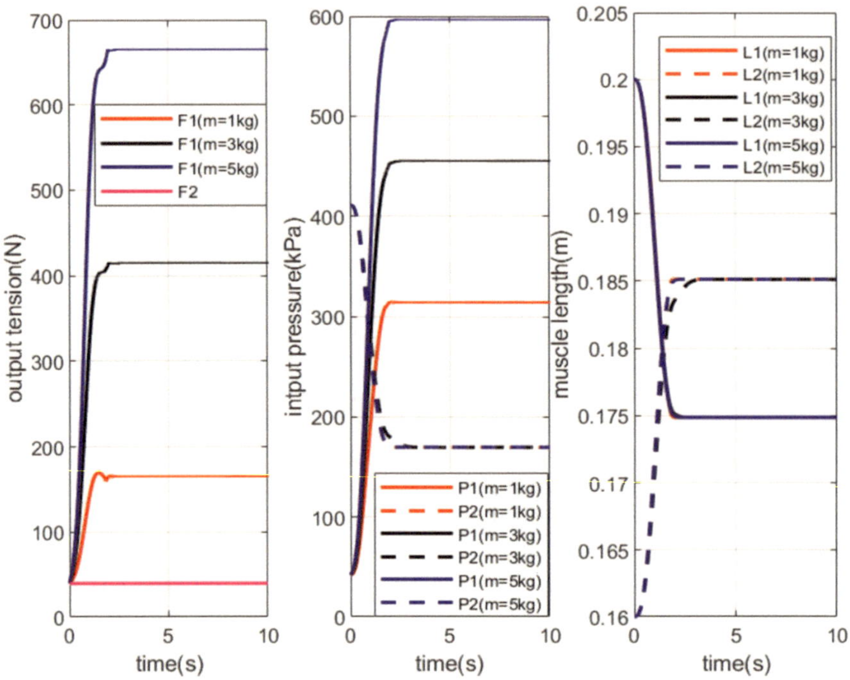

Figure 12.
Simulation result of the tension, pressure, and length of the PAMs with different payloads.

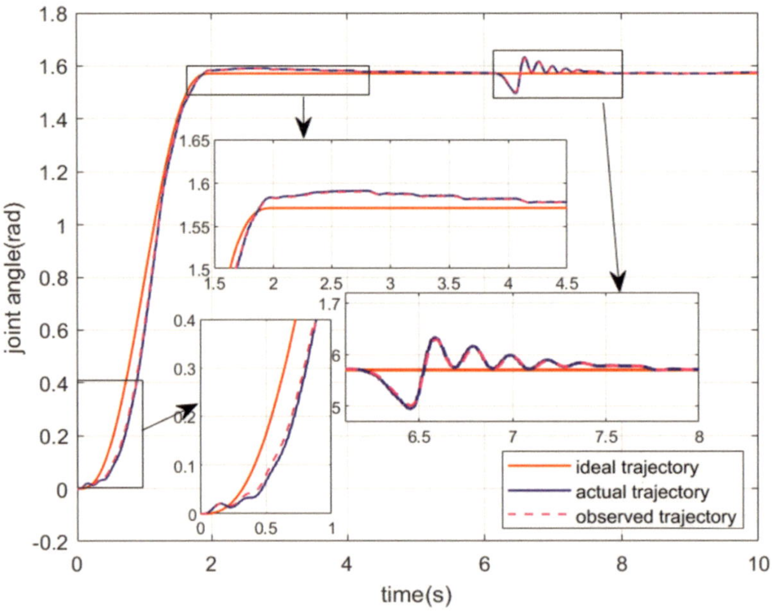

Figure 13.
Physical experiment result of the angular trajectory of the robotic joint ($z_3(0) = 130$).

3.5 Summary

It is generally acknowledged that it is very difficult to model PAM precisely, and the model imprecision, which, caused by both nonlinearities (including hysteresis effects) and the uncertainty of parameters, will severely impede the control performance of the PAM system. The fSMC-ESO approach views the model imprecision as a kind of internal disturbance, observes both the external and internal disturbances via ESO, and compensates them via fSMC to realize the human-like motion with expected robustness and tracking accuracy. As validated in this experiment, though the hysteresis effects of the PAM are completely ignored by taking only one part of Yu's model (these two parts obviously different from each other) into consideration, this approach still is able to achieve good control performance of the robotic joint system. Therefore, the fSMC-ESO approach can be applied to solve the model imprecision problems of the PAM systems and consequently improve their control performance.

4. Limitations and relevant promising studies of PAM systems

PAMs do possess some distinct advantages, which are favorable for the actuation of the cooperative robots; it also has some unfavorable properties except the strong nonlinearities and the uncertainty of parameters that can be properly handled as abovementioned. These topics are briefly discussed as follows:

4.1 The pressured air source issue

PAM works with the compressed air as mediator, and via the outlet valve, it vents the high-throughput compressed air to the atmosphere. Therefore, a PAM system requires a pressured air source, e.g., an air compressor with high-pressure gas pot, which undoubtedly will make the PAM system bulky.

A solution of this issue is to make the working medium recyclable. Therefore, researchers studied the Hydraulic McKibben Artificial Muscles (HAMs), which take the hydraulic oil or water instead of pressed air as the working fluid.

The most impressive advantage of the HAMs is its supper high energy efficiency, e.g., a prototype high force HAM, which is 40 mm in diameter and 700 mm in length and driven by a maximum water hydraulic of pressure of 4 MPa, can achieve a maximum contracting force of 28 kN [41]. However, HAMs also degrade its inherent compliance because of the incompressible hydraulic working fluid. Furthermore, HAMs did not solve the bulky size problem of PAMs, because it also needs a reservoir, as well as pumps and valves to supply the pressured working fluid.

4.2 The dilemma of the output torque and the motion range of a joint

As illustrated in both the **Figure 2** and the Eq. (1), the PAMs' contraction length ε equals the arc length on the pulley circumference corresponding to the rotary scope ϑ of the joint. For a given item of PAM, the contraction length is a constant, and the pulley radius should be limited so as to cover the required joint rotation angle with the constant contraction length. Consequently, the joint torque $T = F \cdot r$ is limited, which will restrict the application of PAMs in robotic systems.

One can obtain a higher joint torque either via a higher contraction force F or via a larger pully radius r.

The former method is known as the Variable Recruitment Strategy (VRS), which parallelly deploys many PAMs as a muscle bundle and recruits different number of PAMs according to the output torque requirement [42]. VRS recruits 1, 2, or 3 pairs of PAMs to generated different force for meeting the application requirement, respectively. Consequently, the robotic system is able to realize the expected actuating performance with the minimal activated number of PAMs for increasing the efficiency of the PAM system.

However, the VRS has some drawbacks that impede its robotic application:

1. The PAMs are parallelly deployed to generate a variable output torque via a variable number of recruitments, while both the contraction length and the pully radius of the PAM group configuration are the same as those of a single pair of PAM configuration, the power efficiency of the joint therefore is not high;

2. In the working process, the PAMs did not be activated (still with an amount of air in it) will be bended or buckled, which will generate a certain resistance and degrade the drive performance of the joint. Furthermore, the frequent bending will seriously shorten the life span of PAMs;

3. The group of PAMs must be exactly parallelly deployed as a muscle bundle to draw the same steel wire. It consequently will make the system bulky and impede the deployment and application of the VRS PAMs group in a robotic system.

The latter method sequentially deploys and activates multiple PAMs via a large pully to generate a high output torque. As illustrated in **Figure 14**, in order to drive the joint rotating a certain angle, multiple PAMs are activated one after the other, and the total arc length of the pulley circumference corresponding to the rotary scope of the joint is therefore assigned to the total contraction length of these PAMs rather than those of one PAM. An overrunning clutch is employed to transmit the driving torques of PAMs to the joint axis. As a result, the PAMs generate a high output torque on a larger pully radius than those of the VRS approach.

The second approach is expected to solve the aforementioned problems of the VRS approach and accordingly show the following advantages:

1. Higher power efficiency. Because it generates the driving torque of the joint via a larger pully radius and less energy consumption;

2. Flexible deployment of the PAMs. PAMs of a joint are no longer need to be parallelly deployed as a muscle bundle. Therefore, the volume of a joint can be

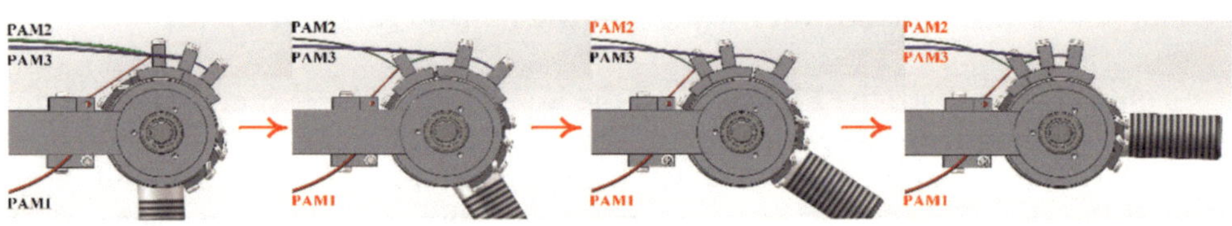

Figure 14.
The sequential activation approach (PAMs in red means be activated).

reduced for convenient of manipulations, and the PAMs will have a longer life span because they will not be bent anymore.

4.3 The flexibility and shape adaptability for soft wearable robot applications

The PAMs are considered a good actuator for wearable robot systems, i.e., the exosuit and the exoskeleton robot, because of its compliance characteristics and the human-muscle-like contraction properties. However, the PAMs commonly used are large in size (10–40 mm in diameter and 100+ mm in length) and rigid in use (drastically stiffen when inflated), which make them unsuitable to be applied to wearable robot applications. The wearable robots require the flexibility and shape adaptability of PAMs for compliantly deforming along with the human body and redundant human-muscle-like actuation mechanism.

A promising study for this problem is the thin McKibben muscle (TMM) or multifilament muscles [43]. When a group of the TMMs with an outer diameter of 1.8 mm are bundled as a robotic actuator, it is flexible (even when pressurized), compact, and lightweight to be easily attached to the complex or irregularly shaped parts of the robots for actuation. Especially, TMMs are deformable to follow the human body shape, as well as taking various shapes to mimic the human muscle shapes, e.g., the biceps, the triceps, and the flat muscles of the pectoralis major or the deltoid muscles, for realizing a human-friendly support and power assistance.

Furthermore, the structural feature of multifilament makes TMM a perfect choice for mimicking both the structure and the recruitment mechanism of the skeletal muscle. Each skeletal muscle has a finite number of motor units, which is the smallest subunit of a muscle that can be recruited and consists of a certain number of muscle fibers 100 μm in diameter. The fiber number of a motor unit varies from 3 (for the fine control of the eyes) to 2000 (for the vigorous motion of the legs) according to its location and function [44]. The human central nervous system will excite (recruit) appropriate number of motor units according the task requirement for the optimization of actuation efficiency. The aforementioned VRS of PAMs is exactly an attempt for this propose, while the multifilament structure of TMM makes it a more suitable choice for imitating the recruitment mechanism of the human skeletal muscles.

TMMs have the same limitation of PAMs, i.e., they require a pressurized fluid source supplied by an external pump or compressor, which are generally rigid and bulky, and consequently preventing untethered and portable operations.

A promising study to this issue is the pumps based on the principle of ElectroHydroDynamics. Cacucciolo et al. studied a solid-state pumping mechanism that directly accelerates the working medium of dielectric liquid by means of an electric field [45]. This pump is silent, fully stretchable, and bendable while operating, which makes it a very promising solution for the TMMs actuated soft wearable robot.

5. Conclusions

The McKibben PAMs possess some compelling properties, e.g., the inherent compliance, the actuating characteristics similar to those of the human muscle, the flexibility of deployment, etc. These properties make PAM an excellent actuator for the humanoid robot systems. Focused on the robot application of PAMs, this chapter concerned the following critical techniques:

Firstly, this chapter presented a solution for the PAM system design. People have tried hard to realize the human-like dexterous operational capability via highly biomimetic design that mimics the musculoskeletal structure of human body in the design of robot system, but inevitably lead to high complexity in structure and difficulty for manufacture and control. In order to solve this problem, this chapter in detail presented the bionic design of a 7 DOF human-arm-like manipulator. This design mimics the human arm in functionality rather than in construction. It takes the antagonized PAMs and Bowden cables that mimic the muscle-tendon-ligament structure of human body and elaborately configures the DOFs of the arm and flexibly deploys the routing of Bowden cables. As a result, the joints of the analog shoulder, elbow, and wrist of the robotic arm intersect at a point respectively though these DOFs are sequentially arranged, and the motions of these joints are independent from each other for convenience of motion control and human-like dexterous manipulation. If motors instead of PAMs are taken as the actuators of the robotic arm, it will be very difficult to make the joints of the analog shoulder, elbow, and wrist intersect at a point respectively, and the robotic arm will be offset and bulky because of the volume of the motors and reduction gears.

Secondly, this chapter presented a solution for the tracking motion control of the PAM system. It is universally acknowledged that the PAM is very difficult for modeling and control because of its strong nonlinear characteristics. This chapter proposed a fSMC-ESO approach to realize the human-like motion control of the robotic joint. This approach views the model imprecision caused by the strong nonlinearity of the PAMs as a kind of internal disturbance of the robotic system and combines ESO with fSMC to observe and compensate the external and internal disturbances. Experiments validated that the fSMC-ESO approach is able to realize the human-like motion with expected robustness and tracking accuracy, though the model of PAM is obviously inaccurate because only one part of the Yu's model is taken into consideration. Therefore, it can be concluded that the fSMC-ESO approach has remarkable advantages for solving the model imprecision problems and achieving good control performance for the PAM systems.

Thirdly, some variants of PAMs that are aiming to amend or remedy the drawbacks of the PAM systems are discussed. Especially, a Sequential Activation approach that targets to the dilemma of the output torque and the motion range of a PAM actuated joint is proposed. Different from the others, this approach sequentially deploys and activates multiple PAMs via a large pully and an overrunning clutch to generate a high output torque at the same time cover the total rotary scope of the joint, and it may also bring advantages of higher power efficiency and more flexible deployment of the PAM systems.

As a conclusion, the McKibben PAM actuator is a promising solution for the advanced robotic systems, and the PAM-based inherent safety and compliant manipulation will certainly play an important role in the cooperative humanoid robots. With further research, people can optimistically anticipate the wide applications of the PAM actuated robotic systems in the human living environment.

Acknowledgements

This work was supported by the National Natural Science Foundation of China (grant number: 61673003) and the Natural Science Foundation of Beijing Municipality (grant number: 4192010).

Conflict of interest

The authors declare no conflict of interest.

Author details

Daoxiong Gong* and Jianjun Yu
Faculty of Information Technology, Beijing University of Technology, Beijing, China

*Address all correspondence to: gongdx@bjut.edu.cn

References

[1] Ding H, Yang X, Zheng N, et al. Tri-Co Robot: A Chinese robotic research initiative for enhanced robot interaction capabilities. National Science Review. 2017;**5**:799-801. DOI: 10.1093/nsr/nwx148

[2] Daerden F, Lefeber D. Pneumatic artificial muscles: Actuators for robotics and automation. European Journal of Mechanical and Environmental Engineering. 2002;**47**(1):10-21

[3] Andrikopoulos G, Nikolakopoulos G, Manesis S. A survey on applications of pneumatic artificial muscles. In: The 19th Mediterranean Conference on Control and Automation; 20-23 June 2011; Greece. New York: IEEE; 2011. pp. 1439-1446

[4] Klute K, Czerniecki J, Hannaford B. Artificial muscles: Actuators for biorobotic systems. The International Journal of Robotics Research. 2002; **21**(4):295-309

[5] Airic's arm—Festo [Internet]. Available from: https://www.festo.com/net/SupportPortal/Files/42058/Airics_arm_en.pdf [Accessed: November 01, 2021]

[6] Robotic Hands That Mimic Human Movement with Air Muscles [Internet]. Available from: https://www.techeblog.com/feature-robotic-hands-that-mimic-human-movement-with-air-muscles/ [Accessed: November 01, 2021]

[7] Nanayakkara V, Cotugno G, Vitzilaios V, et al. The role of morphology of the thumb in anthropomorphic grasping: A review. Frontiers in Mechanical Engineering. 2017;**3**(5):1-21. DOI: 10.3389/fmech.2017.00005

[8] Xu Z, Todorov E. Design of a highly biomimetic anthropomorphic robotic hand towards artificial limb regeneration. In: The 2016 IEEE International Conference on Robotics and Automation (ICRA); 16-21 May 2016; Stockholm. New York: IEEE; 2016. pp. 3485-3492

[9] Faudzi A, Ooga J, Goto T, et al. Index finger of a human-like robotic hand using thin soft muscles. IEEE Robotic and Automation Letters. 2018;**3**(1):92-99

[10] Tasi B, Koller M, Cserey G. Design of the anatomically correct. Biomechatronic the Hand. 2019. Available from: https://arxiv.org/ftp/arxiv/papers/1909/1909.07966.pdf [Accessed: 2021-12-01]. DOI: arXiv: 1909.07966

[11] Ikemoto S, Kannou F, Hosoda K. Humanlike shoulder complex for musculoskeletal robot arms. In: 2012 IEEE/RSJ International Conference on Intelligent Robots and Systems; 7-12 October 2012; Vilamoura. New York: IEEE; 2012. pp. 4892-4897

[12] Andrikopoulos G, Nikolakopoulos G. Design, development and control of a human-inspired two-arm robot via pneumatic artificial muscles. In: 2017 25th Mediterranean Conference on Control and Automation (MED); July 2017; Valletta. New York: IEEE; 2017. pp. 241-246

[13] Andrikopoulos G, Nikolakopoulos G, Kominiak D, Unander-Scharin A, Towards the development of a novel upper-body pneumatic humanoid: Design and implementation. 2016 European Control Conference (ECC); 2016; Aalborg. New York: IEEE; 2016. pp. 395-400

[14] Lilly J, Liang Y. Sliding mode tracking for pneumatic muscle actuators in opposing pair configuration. IEEE Transactions on Control Systems Technology. 2005;**13**:550-558

[15] Estrada A, Plestan F. Second order sliding mode output feedback control with switching gains—Application to the control of a pneumatic actuator. Journal of the Franklin Institute. 2014;**351**: 2335-2355

[16] Shi G, Shen W. Hybrid control of a parallel platform based on pneumatic artificial muscles combining sliding mode controller and adaptive fuzzy CMAC. Control Engineering Practice. 2013;**21**:76-86

[17] Amar R, Tondu B, Hamerlain M. Experimental study of nonsingular terminal sliding mode controller for robot arm actuated by pneumatic artificial muscles. IFAC Proceedings. 2014;**47**:10113-10118

[18] Rezoug A, Boudoua S, Hamerlain F. Fuzzy logic control for manipulator robot actuated by pneumatic artificial muscles. Journal of Electrical Systems. 2009;**9**:1-6

[19] Büchler D, Calandra R, Schölkopf B, Peters J. Control of musculoskeletal systems using learned dynamics models. IEEE Robotics and Automation Letters. 2018;**3**:3161-3168

[20] Song C, Xie S, Zhou Z, Hu Y. Modeling of pneumatic artificial muscle using a hybrid artificial neural network approach. Mechatronics. 2015;**31**:124-131

[21] Ahn K, Nguyen H. Intelligent switching control of a pneumatic muscle robot arm using learning vector quantization neural network. Mechatronics. 2007;**17**:255-262

[22] Chang M, Liou J, Chen M. T–S fuzzy model-based tracking control of a one-dimensional manipulator actuated by pneumatic artificial muscles. Control Engineering Practice. 2011;**19**:1442-1449

[23] Chandrapal M, Chen X, Wang W, Hann C. Nonparametric control algorithms for a pneumatic artificial muscle. Expert Systems with Applications. 2012;**39**:8636-8644

[24] Sadrfaridpour B, Wang Y. Collaborative assembly in hybrid manufacturing cells: An integrated framework for human–robot interaction. IEEE Transactions on Automation Science and Engineering. 2017;**15**: 1178-1192

[25] Bortot D, Born M, Bengler K. Directly or on detours? how should industrial robots approximate humans? In: The 8th ACM/IEEE International Conference on Human-Robot Interaction; 3–6 March 2013; Tokyo. New York: IEEE; 2013. pp. 89-90

[26] Kato R, Fujita M, Arai T. Development of advanced cellular manufacturing system with human-robot collaboration. In: The 19th International Symposium in Robot and Human Interactive Communication; 13-15 September 2010; Viareggio. New York: IEEE; 2010. pp. 355-360

[27] Zanchettin A, Bascetta L, Rocco P. Achieving humanlike motion: Resolving redundancy for anthropomorphic industrial manipulators. IEEE Robotics & Automation Magazine. 2013;**20**: 131-138

[28] Kulic D, Venture G, Yamane K, et al. Anthropomorphic movement analysis and synthesis: A survey of methods and applications. IEEE Transactions on Robotics. 2016;**32**(4):776-795

[29] Berret B, Jean F. Why don't we move slower? The value of time in the neural control of action. The Journal of Neuroscience. 2016;**36**:1056-1070

[30] Terlemez Ö, Ulbrich S, Mandery C, et al. Master Motor Map (MMM)—Framework and toolkit for capturing, representing, and reproducing human motion on humanoid robots. In: 2014 IEEE-RAS International Conference on Humanoid Robots; Madrid. New York: IEEE; 2014. pp. 894-901

[31] Gong D, He R, Wang Y, Yu J. Bionic design of a 7-DOF human-arm-like manipulator actuated by antagonized pneumatic artificial muscles. In: The 9th IEEE International Conference on CYBER Technology in Automation, Control, and Intelligent Systems; 29 July 29—2 August 2019; Suzhou. New York: IEEE; 2020. pp. 1503-1508

[32] Gong D, Hao L, Yu J, Zuo G. Bionic design of a dexterous anthropomorphic hand actuated by antagonistic PAMs. In: The 2020 IEEE International Conference on Real-time Computing and Robotics; 28-29 September 2020. Asahikawa. New York: IEEE; 2020. pp. 493-498

[33] Gong D, He R, Yu J, et al. A pneumatic tactile sensor for co-operative robots. Sensors. 2017;**17**:2592. DOI: 10.3390/s17112592

[34] Dahiya RS, Metta G, Valle M, et al. Tactile sensing-From humans to humanoids. IEEE Transactions on Robotics. 2010;**26**(1):1-20. DOI: 10.1109/TRO. 2009.2033627

[35] Gong D, Pei M, He R, Yu J. An extended state observer-based full-order sliding mode control for robotic joint actuated by antagonistic pneumatic artificial muscles. International Journal of Advanced Robotic Systems. 2021:1-11. DOI: 10.1177/1729881420986036

[36] Flash T, Hogan N. The coordination of arm movements: An experimentally confirmed mathematical model. The Journal of Neuroscience. 1985;**5**:1688-1703

[37] Yu H, Guo W, Tan H, et al. Design and control on antagonistic bionic joint driven by pneumatic muscles actuators. Journal of Mechanical Engineering. 2012; **48**:1-9 (in Chinese). DOI: 10.3901/JME. 2012.17.001

[38] Feng Y, Han F, Yu X. Chattering free full-order sliding mode control. Automatica. 2014;**50**:1310-1314

[39] Gao Z. Scaling and bandwidth-parameterization based controller tuning. In: The 2003 American control conference; 4-6 June 2003; Denver. New York: IEEE. 2003;**6**:4989-4996

[40] Huang Y, Xue W. Active disturbance rejection control: Methodology and theoretical analysis. ISA Transactions. 2014;**53**:963-976

[41] Mori M, Suzumori K, Takahashi M, et al. Very high force hydraulic McKibben artificial muscle with a p-phenylene-2, 6-benzobisoxazole cord sleeve. Advanced Robotics. 2010;**24**: 233-254

[42] Meller M, Chipka J, Volkov A, et al. Improving actuation efficiency through variable recruitment hydraulic McKibben muscles: Modeling, orderly recruitment control, and experiments. Bioinspiration & Biomimetics. 2016;**11**: 065004. DOI: 10.1088/1748-3190/11/6/065004

[43] Kurumaya S, Nabae H, Endo G, et al. Design of thin McKibben muscle and

multifilament structure. Sensors and Actuators A. 2017;**261**:66-74

[44] Winter DA. Biomechanics and Motor Control of Human Movement. 4th ed. Hoboken: John Wiley & Sons; 2009. DOI: 10.1002/9780470549148

[45] Cacucciolo V, Shintake J, Kuwajima Y, et al. Stretchable pumps for soft machines. Nature. 2019;**572**:516-519. DOI: 10.1038/s41586-019-1479-6

Section 3

Exoskeletons and Actuators

Chapter 4

Shape Memory Alloy (SMA)-Based Exoskeletons for Upper Limb Rehabilitation

Dorin Copaci, Janeth Arias, Luis Moreno and Dolores Blanco

Abstract

This contribution presents the advances in the use of flexible Shape Memory Alloy (SMA)-based actuators for the development of upper limb rehabilitation exoskeletons that have been carried out by our research group. The actuator features developed by our research group maintain the SMA wire characteristics (low-weight, low-cost, noiseless operation, compact, and simplicity) and additionally presents the flexibility and its increase the work frequency. These characteristics make that its integration in rehabilitation exoskeletons provides the user more comfort, easy to use, and freedom of movement. The chapter describes some different rigid and soft rehabilitation exoskeletons for different joints such as the elbow, wrist, and hand in which this type of actuator has been successfully integrated. This gives the possibilities to expand the research line with the actuated soft exosuits systems, in a future development perspective.

Keywords: shape memory alloy, rehabilitation devices, elbow exoskeleton, wrist exoskeleton, soft exo-glove

1. Introduction

According to the World Stroke Organization (WSO) [1], almost 14 million people have their first stroke every year, and worldwide over 80 million people are living with the impact of stroke or cerebrovascular accident (CVA). Additionally, researchers have estimated that, as of 2019, there are more than 17,000 new cases of SCI (spinal cord injury) each year and between 249,000 and 363,000 people are currently living with this injury in the United States [2]. These types of disorders, in most of the cases, are associated with the partial or total loss of the sensory motor and autonomic function. The persons affected by these disorders present a lower quality of life and often dependent on other persons. It is possible to recuperate one part of these loosed sensory motor function with the aid of the rehabilitation therapy, but these treatments are very expensive in health resources and very long in time.

Today, the wearable exoskeletons are present in the hospitals and rehabilitation centers, such as support in the rehabilitation therapy. Although most of this rehabilitation devices focused on the lower limb rehabilitation, commercial solutions such Armeo Power from Hocoma [3], InMotion Arm for Neurological Rehabilitation [4],

Amadeo from Tyromotion [5] AlexARm from Kinetek [6] can be founded for the upper limb rehabilitation. Most of these solutions are static devices, with different degrees of freedom (DOF) actuated by DC motors, designed to do the rehabilitation therapy with the patients in the specialized centers. Although the development of the rehabilitation devices for the upper limb was approached in the last years, at present there is still a lack of improvements in this field, so that these devices can be used not only in rehabilitation therapy but also daily life. In this way, the exoskeleton offers the users more autonomy and at the same time improves his quality of life. To optimize the future exoskeletons, different improvements are suggested according to the patient's opinion, which tested these devices. The order proposed by them was easy to use, small and lightweight, tailor-made, safe, comfortable, less distinctive, durable, and affordable [7]. Many of these characteristics are directly related to the actuators used in these devices.

In the past years, the exoskeletons, especially those of the upper limb, were actuated by different types of actuators: DC and AC motors, pneumatic actuators, hydraulic actuators, and other types of actuators such as the shape memory alloy (SMA) [8]. Although the electric motors are one of the most common actuation systems for the exoskeleton, these are still limited by characteristics such the weight, need of gearboxes to reduce the velocity, and the operation noise. On the other hand, the hydraulic and pneumatic actuators present a good force-weight relation but still limited by the noise and the need of compressed air. The Shape Memory Alloy (SMA) is a metallic alloy, which has the property of recovering its original shape (the memorized shape) after being deformed when heated above the transformation temperature between a martensite phase (at low temperature) and an austenite phase (at high temperature). This presents a good force-to-weight ratio, small volume, and noiseless operation, the SMA-based actuators being considered a good actuation solution for wearable and soft robotics applications and in particularly for rehabilitation devices. The principal disadvantages of this type of actuators are represented by the hysteresis effect, which makes its control difficult, and the low work frequency. These disadvantages limit the use of this type of actuators for certain applications.

Recently, this type of material was used as an actuator in various rehabilitation devices for lower and upper limb and for prosthesis. In [9], a glove actuated by SMA for rehabilitation exercise and assistance was presented. This soft robotic device can provide for the user in grasping 40 N force. The actuator used in this device is based on an SMA wire with diameter of 0.38 mm, cooled by air fans. In [10], the SMA wires were used as actuator for a 3 DOF wrist rehabilitation device. Similarly, in this work, to improve the cooling stage of the actuator, mini air fans were used. The proposed methods do not present the actuator flexibility, and with the air fans, the size of the device increases. In [11], the SMA was used as a hybrid actuator for a hand exoskeleton, combining the SMA springs with a servomotor. Also, the SMA springs were used as actuators for a soft wrist assistive device [12]. In [13] three SMA wires were used in parallel configuration as actuator in a suit-type elbow flexion assistance. For the lower limb, the SMA actuator was embedded in smart clothes for the ankle assistance [14]. This is a totally soft device, which can assist in the ankle with a torque of 100 Ncm. In most of the publications, the authors do not give details about the actuator position response on the cooling stage, where the actuator needs to cool to extend. This necessary time depends on the wire's diameter, ambient temperature, and if it is or not forced to cool, and this time can affect the device performances.

Our research group, RoboticsLab from Carlos III University of Madrid, Spain, developed different exoskeletons for the upper limb rehabilitation actuated by

SMA-based actuators. For the user's comfort, we propose lightweight exoskeletons, but maintaining the power performance of a rigid exoskeleton. Also, the proposed devices have noiseless operation, low-cost fabrication, and are more compact. These exoskeleton characteristics, in great measure, are due to the used actuator—a flexible structure based on Bowden cable without additional cooling system. According to the proposed actuator based on SMA, we developed three different exoskeletons, which will be presented in this study, for the elbow joint, the wrist joint, and hand rehabilitation. Each one presents two or more DOF according to the articulation where it operates, and the actuators have the possibility to work in antagonistic configuration. According to this configuration, the position error decreases significantly in the cooling stage.

This study is divided into four sections. Section 2 presents the proposed SMA-based actuator used in the exoskeleton structure with its electronic hardware and its control algorithm. This section continues with the presentation of the developed exoskeletons from our laboratory, which have used the SMA-based actuator. Section 3 presents the discussions in terms of the current and future perspective of rehabilitation exoskeleton improvements. Section 4 introduces some conclusions and future works.

2. Methods

This section presents the SMA-based actuator used on the upper limb exoskeletons with its electronic hardware and its control algorithm. Also, in this section, the different exoskeletons configurations for the upper limb, elbow, wrist, and hand, will be presented.

2.1 Shape memory alloy: based actuator

The actuator used in rehabilitation devices is based on SMA and consists of one or more SMA wires, a Bowden cable, a polytetrafluoroethylene (PTFE) tube, and the terminal parts. The actuator force and its dimensions can vary depending on the number of wires and their diameter. According to the necessary force to mobilize different upper limb joint, three wire diameters was considered. The characteristics of these SMA wires used in the configuration of different actuators can be seen in **Table 1**, where the current represents the approximate current for 1 second contraction.

The actuator structure with a single SMA wire can been seen in **Figure 1**, left side. On the right side, a schematic actuator cross section can be observed. The actuator has been adapted in length, diameter, and number of wires according to the final application. The principal components of the actuator, enumerated in **Figure 1**, are detailed below:

Diameter size	Resistance	Current	Force	Cooling 70°C	Cooling 90°C
(mm)	(Ω/m)	(A)	(N)	(s)	(s)
0.31	12.20	1.50	12.55	8.10	6.80
0.38	8.30	2.25	22.06	10.50	8.80
0.51	4.30	4.00	34.91	16.80	14.00

Table 1.
Properties of the SMA wires [15].

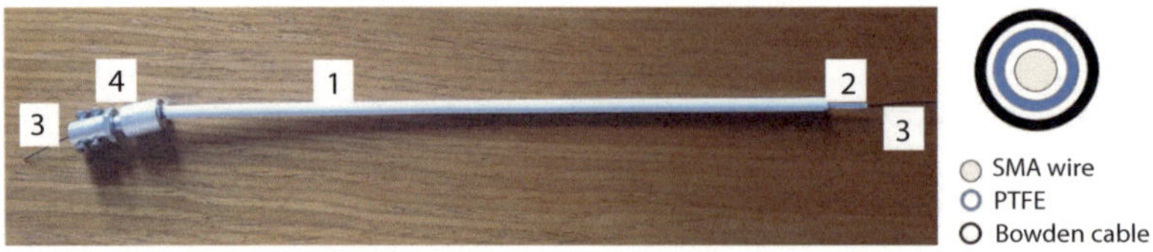

Figure 1.
SMA-based actuator. Right side: 1 – Bowden cable; 2 – PTFE tube; 3 – SMA wire; 4 – Terminal unit; left side, actuator cross section.

- 1 – Bowden cable. It is a type of flexible cable used to transmit the force. In this case, it is composed of a metallic spiral covered with a nylon sheath. This gives the flexibility advantage of the actuator and helps to dissipate the heat when the SMA wire is in the cooling stage (recovering the initial length). In **Figure 1**, a Bowden cable with 3.5 mm diameter is represented. This Bowden cable is used only for actuators with only one SMA wire. For actuators with more SMA wires, a Bowden cable with diameter 6.5 mm is used. In this last case, depending on the SMA wire diameter, the actuator can have up to five wires if the SMA wires have a diameter 0.51 mm.

- 2 – PTFE tube. It is transparent, chemically inert, and nontoxic material, which facilitates the SMA wires displacement, considered to be a solid lubricant. This is placed between the SMA wire (or the SMA wires for the multi wires actuator) and the Bowden cable, acting as an electrical insulator. In addition, it can also work at high temperatures, over 250°C.

- 3 – SMA wire. In **Figure 1**, the actuator is composed of only one SMA wire. The actuator structure can be modified to include more SMA wires, whose diameter and length are calculated according to the necessary force and the final displacement of the device.

- 4 – Terminal unit. This is used to fix the SMA wire with the Bowden cable, at one end, and the SMA wire with the actuated system or the tendons of the actuated system, at the opposite end. The terminal unit is composed of two pieces screwed together, which permit to tense the SMA wire, after being mounted in the final application. Furthermore, those terminal units are used as connectors for power supplying the actuator.

In the exoskeleton structures presented in this chapter, the multi-wire actuators have all the SMA wires inside of only one PTFE tube and everything in a Bowden cable.

This flexible SMA actuator based on Bowden transmission system certainly has some features that make it a good alternative to the use of conventional actuators in soft exoskeletons. Using long SMA wires inside a flexible tube makes it possible to design an actuator that can provide the necessary displacements required by soft exoskeletons. Also, these are easy to integrate and adapt into the flexible and dynamic structures. The possibility of flexing and physical arrangement of the actuator in almost any way has allowed us to better approach the "soft-robotics" concept, so that the actuator no longer imposes rigid mechanical structures on the joints [16].

2.1.1 Electronics hardware

The electronic hardware consists of one or more position sensors depending of the rehabilitation device (these will be detailed when each device will be described), a microcontroller, and a power circuit required to control the SMA-based actuators.

The electronic power circuit for SMA wires is based on MOSFET transistors. The transistors are activated by pulse width modulation (PWM) provided by the controller. The transistors open and close the circuit with a power supply for the actuators. With these electronics (developed by our research group), the control hardware architecture can manage two, four, or six different actuators (each actuator with one or more SMA wires).

The controller board is based on the STM32F407 Discovery kit [17], from STMicroelectronics, which is programmed with Matlab/Simulink [18]. This manages signals from the sensors, executes the control algorithm for controlling the actuators, and generates the required PWM signals.

2.1.2 Control algorithm

Due to the characteristic of hysteresis and the nonlinear behavior of the SMA-based actuator, the control algorithm is a quite complex. A bilinear proportional integral derivative (BPID) controller was proposed to compensate these nonlinearities, which schematically is presented in **Figure 2**. This is based on previous works and the literature [19–21].

In **Figure 2**, the BPID controller is schematically represented where: Y_{ref} represents the desired reference, V is the control signal generated by the proportional integral derivative (PID) controller, U represents the control signal rectified by the bilinear term, and Y represents the actuator position response. The performance of this control algorithm was previously tested and compared with other two controllers, a conventional PID and a commuted feedforward PIPD, controlling a real SMA actuator [19].

2.2 Exoskeletons for upper limb rehabilitation

In this section, different exoskeletons prototype developed by our research group is presented. According to the target joint (elbow, wrist, or hand/fingers), the proposed actuator is implemented in different configurations: with only one or more wires with different diameters and lengths.

2.2.1 Elbow exoskeleton

The elbow joint is a complex articulation that helps to position the hand in space. The humeroulnar and the humeroradial articulations are classified as hinged joints and permit the elbow flexion extension movement. On the other hand, the proximal

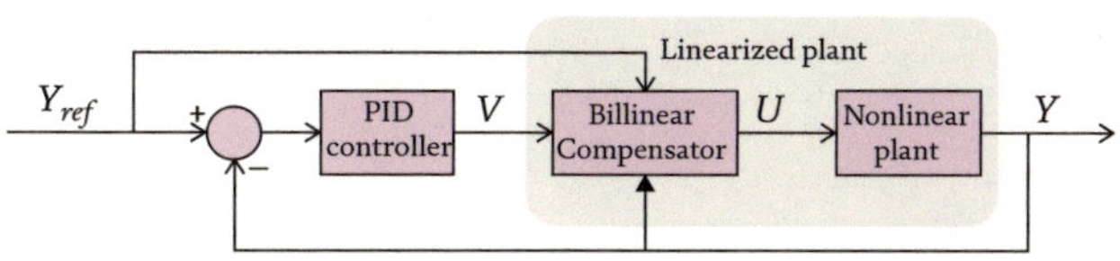

Figure 2.
BPID control algorithm.

radioulnar articulation permits the forearm pronation and supination movement and is classified as a trochoid joint [22]. Although the elbow joint in the flexion-extension movement permits a range of movement between 0 and 150 degrees, in a daily living (ADL), the functional range is estimated between 30 and 120 degrees. Similarly, the human body permits approximately 71 degrees of pronation and 81 degrees for supination, though in the ADL the functional range is estimated in 50 degrees of pronation and 50 degrees of supination.

The proposed device can be seen over the human body in **Figure 3** (left side frontal plane and right side sagittal plane) and was detailed in a previous work" SMA Based Elbow Exoskeleton for Rehabilitation Therapy and Patient Evaluation" [23]. This has two degrees of freedom (DOF), which permit the movement of flexion-extension and pronation-supination. For safety, the flexion-extension movement was mechanically restricted between 0 and 150 degrees and the pronation-supination movement between −60 and 60 degrees. This is a low-cost device with most of the pieces 3D printed except the pieces that are subjected to high forces made in aluminum. Although it has a rigid structure, this can be set according to the patient segments (arm and forearm) dimensions to maintain the exoskeleton rotation axis aligned with the biomechanics of human body (elbow axis). This can be easy set customizing the exoskeleton for each patient. The segments and articulation of the device are mechanically restricted according to the human body limitations, to carry out a safe rehabilitation therapy. Due to the SMA-based actuator, the exoskeleton presents a noiseless operation and more compact dimensions, which make it less distinctive. The total weight of this device including the actuators is less than 1 kg, which can be classified between the most lightweight elbow rehabilitation devices with 2 DOF.

The actuators used in this device are based on the SMA wire with 0.51 mm of diameter. The actuators for the flexion-extension movement are composed of four SMA wires each in the same PTFE tube and a Bowden cable, as presented in the Section 2.1. Each actuator in this configuration can exert a nominal force of approximately 140 N, and considering that the linear displacement is converted to rotary displacement through a pulley with a diameter of 0.06 m, the nominal torque in the elbow exoskeleton joint is around 4.2 Nm (a maximum torque of 13.56 Nm). These two actuators work in antagonist configuration, simulating the biceps–triceps muscle

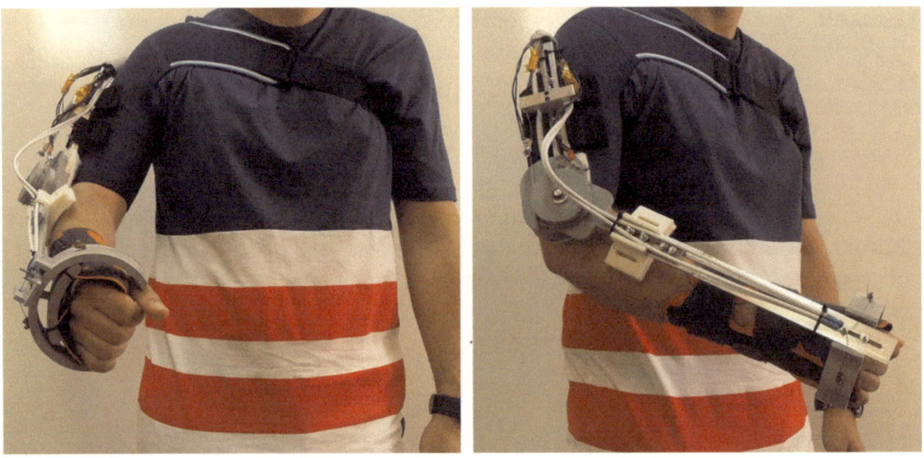

Figure 3.
Elbow exoskeleton over the human body.

group. For the prono-supination movement, the actuators each are based only one SMA wire, each one presenting a force of 35 N. According to the necessary displacement, the actuators have a length of 1.5 m for the flexion-extension and 2 m for the prono-supination. The total weight of the actuators is around 0.54 kg.

The exoskeleton was tested and evaluated with the healthy subjects and post-stroke patients. In total 10 patients with age 61.8 ± 12.98 and six physiotherapists tested the elbow joint exoskeleton and completed the usability test, QUEST 2.0 [24]. The test results were promising with a score of 33 ± 6.90, where the most appreciated items were the weight and dimensions of the exoskeleton, both scored 4.3 ± 0.674. The least appreciated was the item of effectiveness scored with only 3.8 ± 1.03, followed by the comfort and simplicity. These results were influenced by the fact that during the tests, the exoskeleton was in an improvement stage and only was tested in passive mode where the patients with the activity in the motor function do not consider it useful for their rehabilitation therapy.

An active rehabilitation therapy, with the elbow exoskeleton, based on the superficial electromyography (sEMG) signals from the biceps–triceps muscles groups was proposed in [25]. The position reference trajectory for the elbow exoskeleton was generated according to the user movement intention detected on the sEMG signals. This approach improves the exoskeleton effectiveness due that the user is motivated to participate in rehabilitation therapy. The elbow exoskeleton response according to the position reference generated in accordance with the sEMG signals can be seen in **Figure 4**. Here the blue signal represents the position reference generated by the high-level control algorithm, and the red signal represents the exoskeleton angular position. The green signal represents the normalized sEMG signals from the bicep muscle. The first $tc = 20$ seconds, the user does a flexion extension movement to calibrate the algorithm and after that the actuator is activated. In this test, the exoskeleton is placed over the subject and only the flexor actuator is used. The exoskeleton is capable of accurately following the reference signal on the flexion movement from 20 to 100 degrees. In extension movement, for example, the interval $t = 100$ to $t = 120$ seconds, the position error increases due to the necessary actuator cooling time. This error can

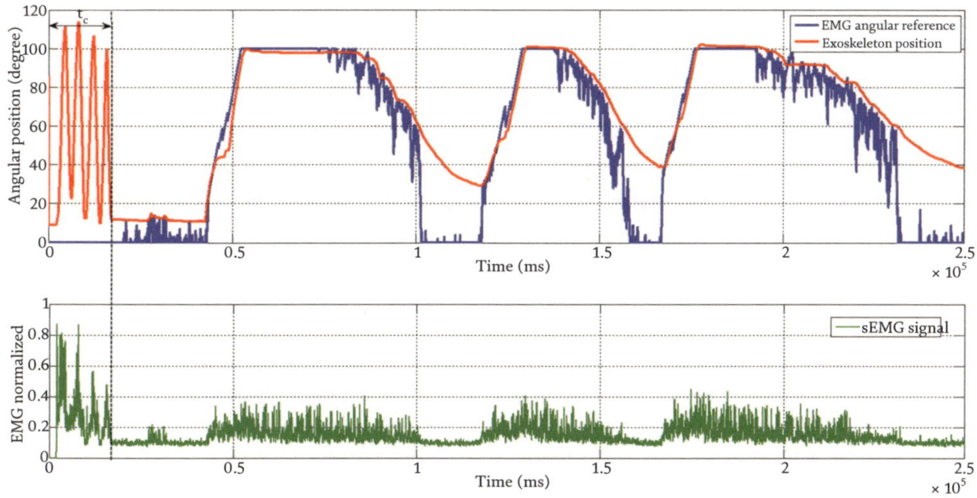

Figure 4.
Elbow exoskeleton position response according to the sEMG signal activation.

be minimized using an antagonistic actuators configuration, similar to that presented for the wrist exoskeleton (Section 2.2.2).

2.2.2 Wrist exoskeleton

The wrist or carpus is a collection of bones, ligaments, tendons and soft tissues, which connect the forearm with the hand. This complex structure offers a wide range of movement that increases the function of the hand and fingers while also giving them a considerable degree of stability [22]. The wrist articulation plays an important role on the daily life manipulation tasks because its kinematic function allows the orientation of the hand with respect to the forearm, and the kinetics allow the transfer of loads from the forearm to the hand and vice versa. The wrist is composed of several joints that make the connections between the radius and ulna bones with the metacarpal bones and the connections with the first and second row of the carpal bones (midcarpal). The wrist joint presents two movements: in the sagittal plane, presents the flexion-extension movement (90 degrees of flexion and 85 degrees of extension) and in frontal plane, presents the ulnar and radial deviation (ulnar deviation 45 degrees and radial deviation 20 degrees).

The wrist exoskeleton actuated by SMA, proposed by our research group, can be seen in **Figure 5** [26]. This presents 2 DOF, one for the flexion-extension movement and the second one for the radial deviation and ulnar deviation. The range of movement achieved with this rehabilitation device is 15 degrees for the flexion, 35 degrees for the extension, 15 degrees with the radial deviation, and 20 degrees with the ulnar deviation. A large part of the device structure is 3D printed and together with the actuators and electronic hardware weighing less than 1 kg. Similar with the elbow joint exoskeleton and the hand rehabilitation glove, due to the actuators' properties, this is considered a lightweight rehabilitation device with a noiseless operation.

The actuators of this device are based on SMA wires with 0.51 mm of diameter and are composed of only one SMA wire, inside the PTFE tube and everything inside the Bowden cable. According to the necessary displacements for the wrist mobilization, and according to the electronic power supply, all the actuators of this device present 2.2 m length. With these characteristics, the rehabilitation device can generate a

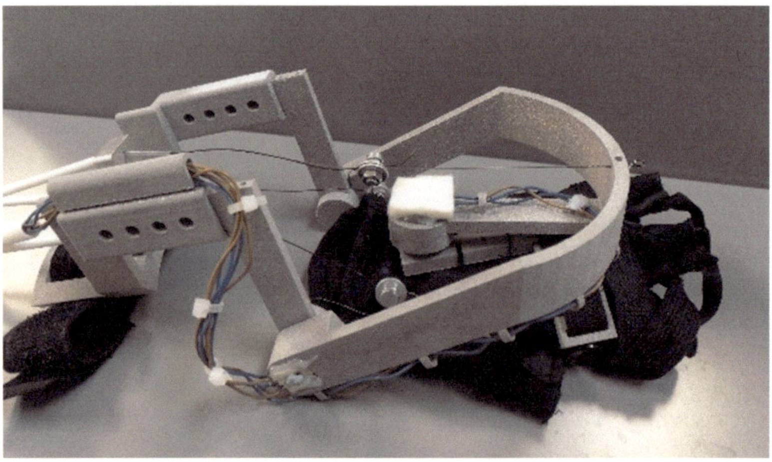

Figure 5.
Wrist exoskeleton actuated by SMA.

torque greater than 0.5 Nm in the wrist joint. The length of the actuators does not represent an inconvenience, considering their flexibility and the possibility to adapt to the shape of the human body.

Considering that during the rehabilitation therapy, the movements are slow, and continuous, a possible reference can be the sinusoidal one. For example, the step reference is not considered because a sudden movement can cause a muscle spasm. **Figure 6** presents the wrist exoskeleton position response on the radial-ulnar deviation with a healthy subject. The control strategy used in this test was based on BPID controller in an antagonist configuration. This configuration works similar such the flexor–extensor muscles group: when the flexor muscles contract the extensors relax and vice versa. In this device, the actuator for the radial deviation was mounted in an antagonist configuration with the radial deviation actuator. The advantage of this configuration consists of decreasing the position error generated by the SMA, the necessary time in a cooling stage to recuperate the initial shape (when it was cool) and by the hysteresis effect. The disadvantage of the antagonist configuration is that after some cycles of continuous work, both actuators present a high temperature, and the system needs to stop to avoid the SMA wires breakage [23].

In **Figure 6**, the actuators flowing a sinusoidal reference with one cycle each 25 seconds. The wrist exoskeleton presents three degrees of error, and the device works continuously during 150 seconds. The work frequency of this actuator is not a problem considering that the rehabilitation device is proposed for the first stage of rehabilitation where the movements are slowly. On the other hand, the number of cycles of continuous work in this case was 6, one cycle every 25 seconds. Although, after 150 seconds, the system was forced to stop, the device can alternate with the flexion-extension rehabilitation for a continuous rehabilitation therapy.

The proposed device has considered improvements compared with the current solutions such as portability, noiseless operation, low cost of fabrication, comfort,

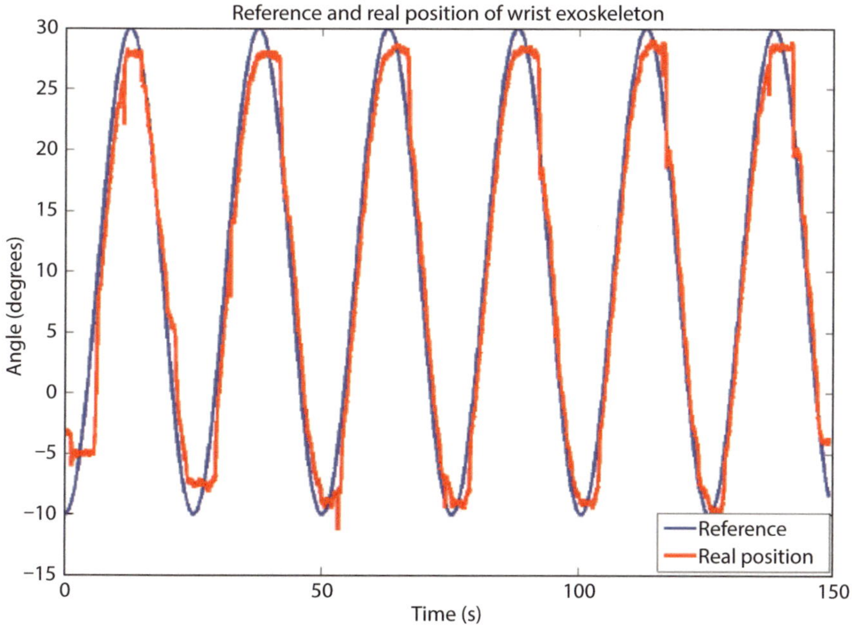

Figure 6.
Position of wrist exoskeleton for radial-ulnar deviation [26].

safety, and easy installation, largely due to the used actuator. The main disadvantage of this device is represented by the slow work frequency, which makes the system only viable for slow rehabilitation therapies. Also, this obligates the system to alternate the therapy between the flexion-extension movement and radial-ulnar deviation.

2.2.3 Soft exo-glove

Hand function plays a fundamental role in performing ADL, maintaining an independent and healthy quality of life. When stroke, SCI, or different neuromuscular disorders occurs, and the hand is affected, the quality of life decreases, and the affected person even becomes dependent on another person. The human hand is a highly complex and multifaceted mobile effector organ that allows it to grasp and manipulate objects. The thumb together with the fingers permits us to manipulate different small objects during daily tasks. Each finger is composed of one metacarpal and three phalanges, and the thumb is composed of one metacarpal and two phalanges, which make that the hand has in total 27 DOF.

In **Figure 7**, a soft exo-glove developed by our research group can be seen. This is actuated by 12 actuators based on SMA wires in antagonistic configuration: six for the fingers flexion and six for the fingers extension. Each group of six actuators is divided into: one actuator for each finger and two actuators for the thumb (these two actuators permit complex movements such as thumb opposition). The SMA-based actuators are connected to the actuation box, where the position sensors are, and where the connection between the actuators and tendons is done. The tendons are routed and fixed over the glove, where its routing represents the key for the realization of the desired movement when the actuators are activated.

The actuators of this device are based on SMA wires, with diameter of 0.38 mm, which presents a force of 22.06 N. According to its characteristics, this can cool after contraction in approximately 8.8 seconds. Considering that the tendon displacement with the proposed routing is around 0.07 m and the SMA actuator when activated contracts 4% of its total length, the total length of each actuator is 2 m. Due to the actuator flexibility, this can take the arm shape and easily can be collocated behind the user.

The developed rehabilitation device is considered totally soft, except the sensors box (where also the connection between the actuators and tendons is done). The actuators, as well as in the other devices (elbow exoskeleton and wrist exoskeleton), are not in contact with the human body, found in the PTFE tube, inside in a Bowden tube, and everything in a flexible PVC tube. With this configuration, the temperature of the actuators is not felt by the user [23].

The future works of this research will focus on integrate the Myo Armband sensor [27] for the hand gesture recognition from the superficial electromyography (sEMG)

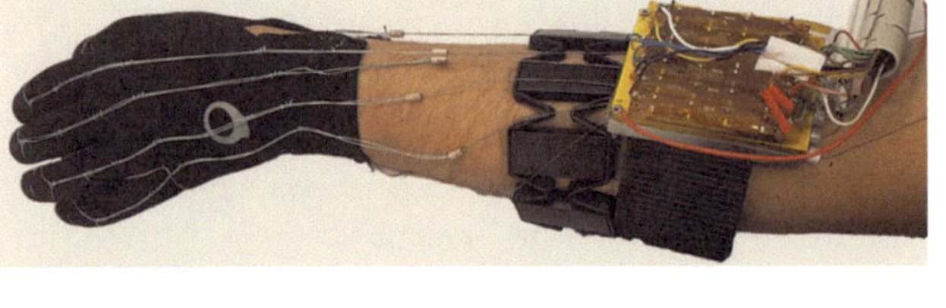

Figure 7.
Soft exo-glove for rehabilitation therapies.

signals. This gives the possibility to realize the active rehabilitation therapies, according to the user movement intention.

3. Discussions

The exoskeletons used during the daily activities offer to the users/patients more autonomy and reduce their dependence on other persons. Also, this improves users' lives and enhances their perceived well-being and sense of community integration [28]. This perspective to integrate the exoskeletons in the patient's daily life to offer them more autonomy is one of the principal goals currently. This implicates the improvements of the currently wearable rehabilitation devices, strictly following the appropriate procedures according to the physiotherapists feedback. The new wearable rehabilitation structures need to be more easy to use, tailor-made according to the user, small and lightweight, less distinctive and with more autonomy. These characteristics are considered some of the most important topics of improvements and are closely related to the actuation system.

From the future perspective of the wearable exoskeletons, which can be used during the daily life, the actuators need to meet some requirements for safety, simplicity, and lightweight that human–robot interaction requires. For these reasons, recently new actuation solutions are being investigated, among which are the artificial muscles. Solutions such Pneumatic Artificial Muscles (PAM) or Shape Memory Alloy are only some of these examples, being already integrated in some prototypes of rehabilitation device. The force–weight relation makes them an excellent candidate for these devices. However, there are still limitations, in different aspects such as the control, compressed air is needed (in case of PAMs), a low work frequency, and energy efficiency (in case of SMAs). These are only a few current research topics, focused to offer viable solutions for the wearable exoskeleton actuation.

The rigid exoskeletons limit the user's freedom movement, complicating his interaction with the environment in a natural way. According to this, we oriented our development on soft exoskeletons or exosuits, aiming of getting closer to the natural user movement. We try to develop exoskeletons that do not constrain the joints like the rigid structures. For the user comfort, we reduce the external structure weight and the actuator weight but maintaining for the most part the performance of a rigid exoskeleton.

The wearable exoskeletons actuated with the SMA-based actuators, developed by our research group, are accessible, easy to use, lightweight, and compact. The test of these devices with the stroke patients and physiotherapists has presented a great interest, obtaining very positive feedback, which encouraged the exoskeletons development initiative. The most appreciative five items on the elbow exoskeleton evaluation with the test QUEST 2.0 were the weight, dimensions, patient adaptation (ergonomics), and safety. These items are directly related to the actuator proposed and used in these devices. Although these have not yet been tested on patients, the wrist exoskeleton and the soft exo-glove stand out for their small dimensions, lightweight, and ergonomic configuration.

4. Conclusions

This contribution presented the recently work of our research group, RoboticsLab from Carlos III University of Madrid, Spain, in the field of upper limb exoskeletons.

Here were presented three different wearable exoskeletons, for elbow, wrist, and hand rehabilitation, movement of which is produced by the SMA-based actuators. Due to the actuator characteristics and proposed design, these devices present: lightweight, noiseless operation, low cost of fabrication, simplicity, and soft or semi-soft structures. According to these characteristics, the proposed devices are not only rehabilitation exoskeletons, which can be used only in the specialized rehabilitation center, but also have the perspective to be used in daily life.

The proposed SMA-based actuator retains the advantages of SMA wires and, in addition, improves the working frequency and adds flexibility to the actuator. This is a promising solution for different applications and especially for softer exoskeletons, which can better adapt to the patient's requirements and offer better ergonomics. The principal disadvantages of this actuator are the low work frequency (viable for slow movement such as the movements of first phase of rehabilitation therapy) and the energetic efficiency.

The elbow joint exoskeleton was tested with the post-stroke patients and physiotherapists. The items best valued in the QUEST 2.0 test were related in great part with the used actuator: the weight, dimensions, patient adaptation (ergonomics), and safety. Although the wrist and the soft exo-glove have not been tested with patients, these devices also present the same advantages.

The future works will focus on the improvement of the exoskeletons structure, closer to a soft and easy-to-use device, especially improving the current actuation system. Although topics such as the work frequency and efficiency were approached in the previous works [29], these represent the key to develop exoskeletons that can be used like support in daily life, giving a certain autonomy when this is needed.

Acknowledgements

The research leading to these results have received funding from the" Sistema robótico para propiciar la marcha en niños pequeños con Parálisis Cerebral" under Grant PID2019-105110RB-C32/ AEI / 10.13039/501100011033, funded by Agencia Estatal de Investigación (AEI); from RoboCity2030-DIH-CM, Madrid Robotics Digital Innovation Hub, S2018/NMT-4331, funded by Programas de Actividades I&D en la Comunidad de Madrid; and co-funded by Structural Funds of the EU.

Conflict of interest

The authors declare no conflict of interest.

Abbreviations

ADL	Activities of daily living
BPID	Bilineal Proportional Integral Derivative
DOF	Degree of Freedom
PAM	Pneumatic Artificial Muscles
PID	Proportional Integral Derivative
PTFE	Polytetrafluoroethylene
PWM	Pulse width modulation

SCI	Spinal Cord Injuries
sEMG	Superficial electromyography
SMA	Shape Memory Alloy
WSO	World Stroke Organization

Author details

Dorin Copaci*†, Janeth Arias†, Luis Moreno† and Dolores Blanco†
Department of Systems Engineering and Automation, Carlos III University of Madrid, Madrid, Spain

*Address all correspondence to: dcopaci@ing.uc3m.es

† These authors contributed equally.

References

[1] World Stroke Organization. Available from: https://www.world-stroke.org/ [Accessed: November 16, 2021]

[2] Spinal Cord Injury Facts & Statistics. Available from: https://www.sci-info-pages.com/spinal-cord-injury-facts-and-statistics/ [Accessed: November 16, 2021]

[3] Armeo Power - Hocoma. Available from: https://www.hocoma.com/solutions/armeo-power/ [Accessed: November 16, 2021]

[4] InMotion Arm for Neurological Rehabilitation. Available from: https://www.bioniklabs.com/products/inmotion-arm [Accessed: November 16, 2021]

[5] Amadeo- Tyromotion. Available from: https://tyromotion.com/en/products/amadeo/ [Accessed: November 16, 2021]

[6] ALEx Arm. Available from: http://www.wearable-robotics.com/kinetek/ [Accessed: November 16, 2021]

[7] van Dijsseldonk RB, Vriezekolk JE, Keijsers NL, Geurts AC, van Nes IJ. Needs and Wishes for the Future Exoskeleton: An Interview Study Among People With Spinal Cord Injury With Community-based Exoskeleton Experience. 2020. [Accessed: October 18, 2021]

[8] Gopura RARC, Bandara DSV, Kiguchi K, Mann GKI. Developments in hardware systems of active upper-limb exoskeleton robots: A review. Robotics and Autonomous Systems. 2016;**75**(Part B): 203-220. ISSN: 0921-8890. DOI: 10.1016/j.robot.2015.10.001. [Accessed: October 18, 2021]

[9] Hadi A, Alipour K, Kazeminasab S, Elahinia M. ASR glove: A wearable glove for hand assistance and rehabilitation using shape memory alloys. Journal of Intelligent Material Systems and Structures. 2018;**29**(8):1575-1585. DOI: 10.1177/1045389X17742729. [Accessed: October 18, 2021]

[10] Hope J, McDaid A. Development of wearable wrist and forearm exoskeleton with shape memory alloy actuators. Journal of Intelligent and Robotic Systems. 2017;**86**(3–4):397. DOI: 10.1007/s10846-016-0456-7. [Accessed: October 18, 2021]

[11] Yang J, Wei T, Shi H. A novel hybrid actuator for the hand exoskeleton. In: 2021 IEEE 11th Annual International Conference on CYBER Technology in Automation, Control, and Intelligent Systems (CYBER). Jiaxing, China: IEEE – Institute of Electrical and Electronics Engineers; 2021. pp. 271-276. DOI: 10.1109/CYBER53097.2021.9588189. [Accessed: October 18, 2021]

[12] Jeong J, Yasir IB, Han J, Park CH, Bok SK, Kyung KU. Design of shape memory alloy-based soft wearable robot for assisting wrist motion. Applied Sciences. 2019;**9**(19):4025. DOI: 10.3390/app9194025. [Accessed: October 18, 2021]

[13] Park SJ, Park CH. Suit-type wearable robot powered by shape-memory-alloy-based fabric muscle. Scientific Reports. 2019;**9**(1):1-8. DOI: 10.1038/s41598-019-45722-x. [Accessed: October 18, 2021]

[14] Kim C, Kim G, Lee Y, Lee G, Han S, Kang D, et al. Shape memory alloy actuator-embedded smart clothes for ankle assistance. Smart Materials and Structures. 2020;**29**(5):055003. DOI: 10.1088/1361-665X/ab78b5. [Accessed: October 18, 2021]

[15] Dynalloy Inc. - Flexinol Actuator Wire Technical and Design Data. Available from: https://www.dynalloy.com/. [Accessed: 16 November 2021]

[16] Copaci DS, Blanco D, Martin-Clemente A, Moreno L. Flexible shape memory alloy actuators for soft robotics: Modelling and control. International Journal of Advanced Robotic Systems. 2020;**17**(1):1-15. DOI: 10.1177/1729881419886747. [Accessed: October 23, 2021]

[17] Discovery Kit with STM32F407VG MCU. Available from: https://www.st.com/en/evaluation-tools/stm32f4discovery.html [Accessed: October 18, 2021]

[18] Flores Caballero A, Copaci D, Villoslada Peciña Ã, Blanco D, Moreno L. Sistema Avanzado de Protipado Rápido para Control en la Educación en Ingeniería para grupos Multidisciplinares. RIAI Revista Iberoamericana de Automatica e Informatica Industrial. 2016;**13**(3):350-362. DOI: 10.1016/j.riai.2016.05.004. [Accessed: October 18, 2021]

[19] Villoslada Á, Escudero N, Martín F, Flores A, Rivera C, Collado M, et al. Position control of a shape memory alloy actuator using a four-term bilinear PID controller. Sensors and Actuators A: Physical. 2015;**236**:257-272. DOI: 10.1016/j.sna.2015.10.006. [Accessed: October 18, 2021]

[20] Copaci D, Blanco D, Moreno L. Flexible shape-memory alloy-based actuator: Mechanical design optimization according to application. Actuators. 2019;**8**(3):63. DOI: 10.3390/act8030063. [Accessed: October 18, 2021]

[21] Martineau S, Burnham K, Minihan J, Marcroft S, Andrews G, Heeley A. Application of a bilinear PID compensator to an industrial furnace. IFAC Proceedings Volumes. 2002;**35**(1):25-30. DOI: 10.3182/20020721-6-ES-1901.00572. [Accessed: October 18, 2021]

[22] Nordin M, Frankel V. Basic Biomechanics of the Musculoskeletal System. Philadelphia, PA 19103: Lippincott Williams and Wilkins; 2001

[23] Copaci D, Martín F, Moreno L, Blanco D. SMA based elbow exoskeleton for rehabilitation therapy and patient evaluation. IEEE Access. 2019;7:31473-31484. DOI: 10.1109/ACCESS.2019.2902939. [Accessed: October 23, 2021]

[24] Copaci D, Serrano del Cerro D, Alguacil-Diego I, Fernández Vázquez D, Molina-Rueda F, Miangolarra-Page JC, et al. Usability evaluation of SMA based exoskeleton: Pilot testing in post-stroke patients. In: International Conference on NeuroRehabilitation ICNR. Cham: Springer; 2020. pp. 153-157. DOI: 10.1007/978-3-030-70316-5_25. [Accessed: October 23, 2021]

[25] Copaci D, Serrano D, Moreno L, Blanco D. A high-level control algorithm based on sEMG signalling for an elbow joint sma exoskeleton. Sensors. 2018;**18**(8):2522. DOI: 10.3390/s18082522. [Accessed: October 23, 2021]

[26] Serrano D, Copaci D, Moreno L, Blanco D. SMA based wrist exoskeleton for rehabilitation therapy*. In: 2018 IEEE/RSJ International Conference on Intelligent Robots and Systems (IROS). Madrid, Spain: IEEE – Institute of Electrical and Electronics Engineers; 2018. pp. 2318-2323. DOI: 10.1109/IROS.2018.8593987. [Accessed: October 23, 2021]

[27] Huitzil-Velasco I, Pajaro-Cruz JO, Ramírez-Alfaro ID. Test of a Myo

Armband. Revista de Ciencias Ambientales y Recursos Naturales. 2017; **3**(10):48-56

[28] Cahill A, Mc Ginley O, Bertrand C, Lennon O. Gym-based exoskeleton walking: A preliminary exploration of non-ambulatory end-user perspectives. Disability and Health Journal. 2018; **11**(3):478-485. ISSN: 1936-6574. DOI: 10.1016/j.dhjo.2018.01.004. [Accessed: October 23, 2021]

[29] Arias Guadalupe J, Copaci D, Serrano del Cerro D, Moreno L, Blanco D. Efficiency analysis of SMA-based actuators: Possibilities of configuration according to'the application. Actuators. 2021;**10**(3):63. DOI: 10.3390/act10030063. [Accessed: October 23, 2021]

Chapter 5

A Review on Vacuum-Powered Fluidic Actuators in Soft Robotics

Seonggun Joe, Federico Bernabei and Lucia Beccai

Abstract

In the past few years, vacuum-powered soft actuators have shown strong potential due to their promising mechanical performance (i.e., fail-safe, fast response, compactness, robustness, jamming, etc.). Indeed, they have been widely exploited in soft robots, for example, grippers and manipulators, wearable devices, locomotion robots, etc. In contrast to inflatable fluidic actuators, the properties of the materials with which they are built have a stronger influence on the kinematic trajectory. For this reason, understanding, both, the geometry and morphology of the core structure, and the material characteristics, is crucial to achieving the desired kinetics and kinematics. In this work, an overview of vacuum-powered soft fluidic actuators is provided, by classifying them as based on morphological design, origami architecture, and structural instability. A variety of constitutive materials and design principles are described and discussed. Strategies for designing vacuum-powered actuators are outlined from a mechanical perspective. Then the main materials and fabrication processes are described, and the most promising approaches are highlighted. Finally, the open challenges for enabling highly deformable and strong soft vacuum-powered actuation are discussed.

Keywords: vacuum-powered actuation, pneumatic artificial muscle, linear actuation, origami structural instability, morphological design, buckling, hyperelastic materials, 3D printing, soft robotics

1. Introduction

In the recent past, Pneumatic Artificial Muscles (PAMs) have assumed a key role in the implementation of movement in soft robots. The strong research attention is due to their interesting mechanical characteristics for soft actuation (i.e., robustness, high pulling force, high-power to mass ratio, and high energy efficiency), compared to other mechanisms exploiting different actuation sources (i.e., electric motor or tendon driven [1, 2], piezoelectric actuators [3], dielectric elastomer actuators [4, 5], shape memory alloys [6, 7] or polymers [8], and ionic polymer-metal composite actuators [9], etc.). Many researchers and engineers have explored quite a few design principles for PAMs, and a wide range of robotic applications was introduced (i.e., industrial robots [10, 11], wearable [12–14], or medical devices [15–17], etc.) where some of the challenges that rigid-bodied robots cannot overcome were addressed [18–21].

From the material point of view, PAMs are generally built from soft materials that can undergo large deformation under external forces. Softness may result from their intrinsic characteristics (i.e., chemistry) or because of a specific design of their structure [22], or both.

In the design phase, the mechanical performance of PAMs (i.e., deformation ratio and blocking force) can be tuned by optimizing some geometrical parameters to achieve specific morphologies. In particular, a tradeoff between deformability and stiffness should be met, depending on the desired mechanical performances. In the literature, different design principles have been used. As most of such designs show, radial strain is an important parameter given its main influence on the linear deformation ratio of the soft actuator.

Most soft pneumatic actuators operating with positive pressure (also called inflatable fluidic actuators) involve—(i) a section of the actuator that expands with pressure; or (ii) strain limiting components that guide the elastic expansion in the desired direction [23].

In the former case, topological and/or morphological approaches are undertaken to produce radial expansion/contraction by employing specific geometries, such as zigzag patterned pleats or convolutions. For instance, the Peano fluidic actuator consists of a set of tubes arranged side-by-side [24]. At null pressure (atmospheric pressure), the actuator remains completely flat. Once positive pressure is applied, each cylindrical tube is inflated, leading to a vertical contraction. Another example is to exploit a bellow that can reduce radial strain, and enlarge longitudinal strain. This design principle is quite useful since biaxial deformability (contraction and elongation) is enabled upon positive pressure and vacuum [25, 26]. In the latter case (ii) multi-material-based approaches were addressed, such as in McKibben actuators or pleated PAMs (PPAMs). A McKibben actuator generally consists of a cylindrical flexible rubber and/or silicone bladder, sheathed with an inextensible fiber network intersected at a certain angle [27, 28]. Different kinematic trajectories are enabled according to the intersection angle (e.g., extension at the intersection angle of more than 54.7 deg., otherwise contraction) [29, 30]. On the other hand, a PPAM generally consists of a membrane having many pleats, with high tensile stiffness (i.e., woven polyester cloth or Kevlar fabric) [31, 32]. This allows the actuators to contract with dramatic increases in diameter when positive pressure is applied. The pleats are arranged along the longitudinal direction, and as the skin begins to swell, the overall structure undergoes a radial shortening and expansion [33].

As mentioned above, desired mechanical performance for inflatable fluidic actuators can be achieved by adopting proper design principles. Nevertheless, the mechanisms exploiting volumetric expansion are commonly vulnerable, depending on material failures (i.e., delamination, fracture), resulting in degradation of the actuator, and poor reliability. Moreover, given the needed volumetric expansion and omnidirectional deformation, they are more likely to be applied where large spaces are available [23], Vacuum-powered pneumatic artificial muscles (or V-PAMs) represent a promising opportunity in soft actuation and solve some of the issues encountered by inflatable fluidic actuators. They rely on decreasing the volume with vacuum and, generally, they do not expand radially during movement. Rather, they can perform linear movements with a relatively low input vacuum pressure (within few hundred kPa) and avoid a stress concentration that could possibly incur along the overall structure at both local and global levels [34, 35]. These characteristics are pursued to achieve high reliability, and ensure a high bandwidth, allowing to saturate at the desired state with a fast response.

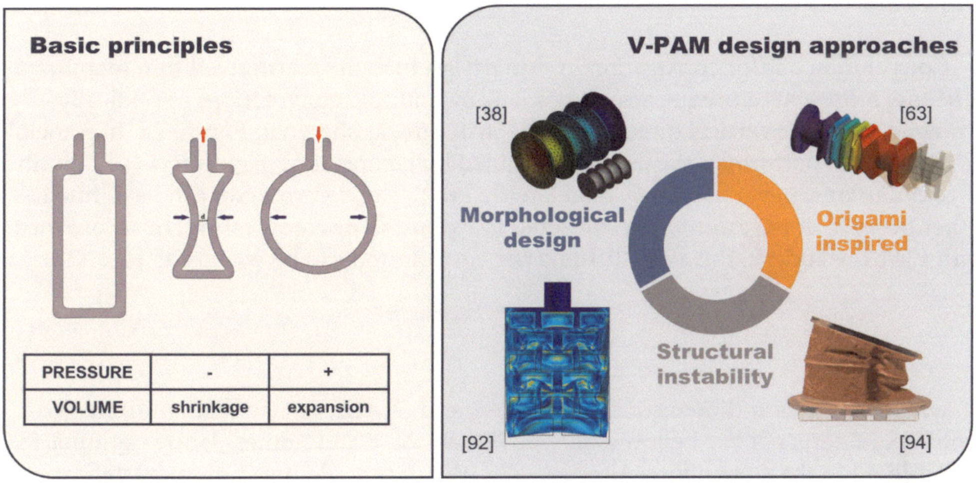

Figure 1.
Left, basic operating principles of pneumatic artificial muscles (PAMs). Right, a classification of vacuum-powered PAMs (V-PAMs) based on the different design approaches: morphological design [38]; origami inspired [63]; structural instability shown in FEM model of the buckling actuator [92], and reproduced with permission [94].

As with inflatable fluidic actuators, the constituting materials strongly influence performance, particularly the deformation ratio. Indeed, due to the incompressibility of the materials, the ratio of bulk modulus (K) to shear modulus (G) is extremely large or becomes almost infinite in case Poisson's ratio is close to 0.5 [36]. Thus, only a slight shortening is achievable until internal walls contact, and collapse, and linear deformation hardly occurs, as shown in **Figure 1**. Moreover, the overall deformations generally include undesired kinematic trajectories (i.e., torsion or bending) due to structural instability (i.e., column squirm and/or buckling). For these reasons, V-PAMs need to be rigorously designed by taking into account geometry and/or morphology, structural stability for imposed deformation and/or load, etc.

Various design principles can be found for V-PAMs in the literature, however (to the authors' knowledge), there are still no comprehensive articles on such an emerging topic, including design, fabrication, and materials. In this work, an overview of the developments of vacuum-powered actuators is provided and they are classified as based on morphological design, origami architecture, and structural instability. The fabrication protocols and soft materials involved are addressed, and their advantages and limitations are discussed and compared.

2. Design

2.1 Morphological design

Soft pneumatic actuators can benefit from designs having zigzag patterns since they enable shear deformation, by enlarging deformations while avoiding structural instability. Typical designs consist of multiple convolutions or corrugations in series along the vertical or horizontal direction of the actuator. In this section, an overview of different types of morphological designs, including actuators with convoluted (or corrugated) skin and pleated skin, is provided by highlighting their pros and cons.

2.1.1 Convoluted skin

Convolution and/or corrugation in skin design includes a crumpled thin membrane. A bellow represents a unique structure made of multiple convolutions periodically arranged along the vertical direction of the structure, as shown in **Figure 2a**. In principle, its mechanical performance can be determined by geometrical parameters (i.e., number of convolutions, mean diameter, total length, etc.). The deformation can be defined as either positive or negative when subjected to extension or compression. In accordance with EJMA[1] standard, the axial stiffness per convolution can be written as [41, 42]:

$$f_{iu} = 1.7\frac{D_m E_b t_p^3 n}{w^3 c_f} \quad (1)$$

where C_f is a non-dimensional parameter, and D_m, E_b, t_p, n, and w, indicate geometric parameters of the bellow skin (see **Figure 2a–c** and nomenclatures summarized in **Table A1** in the Appendix). Alternatively, Wang et al. deduced the axial deformation by means of Castigliano's theorem, and derived the axial stiffness [25], as follows:

$$f_{iu} = \frac{\pi E_b t_p^3}{3N(1-\mu^2)d^2\left[\ln\frac{D}{d} - \left(\frac{D}{d}-1\right) + \frac{\left(\frac{D}{d}-1\right)^2}{2}\right]} \quad (2)$$

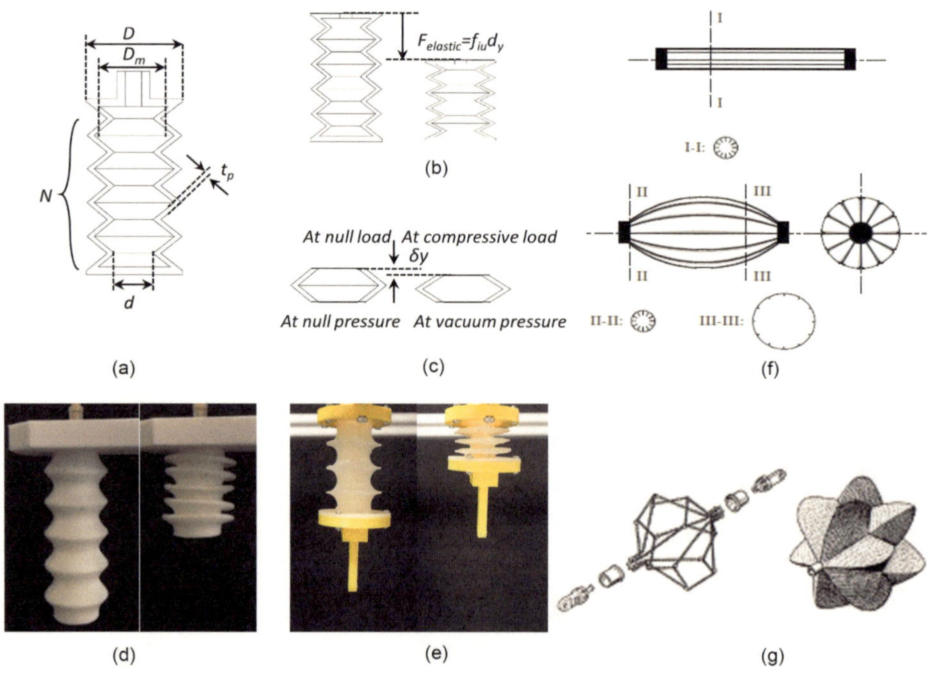

Figure 2.
V-PAMs with morphological design. (a–c) schematic diagrams that show design parameters for a bellow actuator skin. (d) the 3D-printed linear soft vacuum actuator (L-SOVA). Reproduced with permission [37]. (e) the Ultralight Hybrid PAM (UH-PAM) capable of 52% contraction [38]. (f) the Pleated Pneumatic Artificial Muscle (P-PAM) [39]. (g) the Robotic Muscle Actuator (ROMAC) [40].

---

[1] *EJMA: Expansion Joint Manufacturers Association https://www.ejma.org/*

Given Hooke's law ($F_{elastic} = f_{iu}d_y$), the axial stiffness along the y-axis, and the imposed deformation (d_y), identify the elastic force of the bellow structure ($F_{elastic}$). The blocking force of the actuator can be approximated by using a quasi-static analytical model [43, 44]. The work produced by the distributed pressure acting on the inner wall of the skin can be expressed [38] by

$$-F_t \bullet \delta y = P \bullet dv \tag{3}$$

where P and dv are the gauge pressure and the change in the actuator's volume, respectively. Then, the exerted force (also corresponding to the blocking force) at the free end of the actuator can be represented by the derivative of the cross section [45]. The ratio of the outer diameter to the inner diameter (D/d) can be represented as a function of displacement. To the authors' knowledge, it is not possible to approximate the change in volume of actuator explicitly, while it can be obtained by taking into account specific assumptions (e.g., exploiting either geometrical constraints or the material incompressibility) [38]. Indeed, given that the bellows are made of soft or composite materials, they can achieve a linear displacement [46], and it should be highlighted that modeling of the bellow structure is mainly due to the variation in the overall length, rather than radial strain.

Moreover, from Eqs. (1)–(3), it is worth mentioning that the axial stiffness of the bellow is strongly influenced by the skin thickness (t_p) (with cubic order). Also, the higher the number of convolutions, the more compliant the actuator structure is likely to be (i.e., weak stiffness). However, the skin thickness could limit deformability of the bellow at the collapsed state. Thus, both the number of convolutions and skin thickness are important design parameters, determining the axial stiffness and deformability of the overall structure.

Based on such theoretical background, several examples can be found in the literature. Digumarti et al. presented a Hyper-Elastic Bellow (HEB) actuator capable of a euglenoid motion, where both axial and radial strains are enlarged [26]. The bellow, having a half angle of 38.66° with a total length of 50 mm and an outer diameter of 45 mm, was fabricated by using a soft elastomer (Dragon Skin™ 10 SLOW, Smooth-On). The HEB was capable of 40% contraction upon a small vacuum pressure of 8 kPa. More recently, by exploiting additive manufacturing, Tawk et al. presented a 3D-printed Linear Soft Vacuum Actuator (L-SOVA), capable of 51% contraction with a payload of 27.66 N. Another example is to exploit the integration of rigid parts (rings), periodically spaced along the actuator length. This way, Felt et al. showed that a bellow type PAM, made of a tubular thin membrane, can lift a high payload of 21.35 N at −5.52 kPa [47]. Such investigation on V-PAMs has demonstrated that this kind of actuation can enable a high contraction ratio and blocking force, opening the way to new PAM concepts. In refs. [38, 48], our group exploited rigid rings to improve axial stiffness and combined them with open-cell foam modules to build an Ultralight Hybrid PAM (UH-PAM) made of open-cell foam and elastomeric bellow skin (see **Figure 2e**). In this study, we investigated an optimal geometry of the skin and emphasized that even with a light weight of 20 g promising mechanical characteristics can be achieved. Due to the open-cell foam and rigid rings interfaced at each convolution, the axial stiffness of UH-PAM was enhanced, resulting in a high payload of 3 kgf and a high contraction ratio of 52%.

Furthermore, a bellow textile muscle (similar to the Peano fluidic muscle), made of fabric and multiple round discs along its edges and center, was developed to enlarge the deformation ratio [49]. Due to such inflatable and flexible material, the bellow's textile muscle can be displaced to extremely flat (i.e., high contraction ratio of 89%) with a high payload of 32 N at −5 kPa. Recently, Yang et al. extended this work and presented a high-displacement PAM, converting the horizontal motion of the bellow actuator to a vertical motion, based on the contraction when subjected to positive air pressure [45]. Given that the contraction ratio of the bellow is strongly influenced by its thickness, exploiting textile for the skin enables to deform it to a completely flat configuration, upon vacuum pressure. On the other hand, the thin-walled structure could suffer from a lack of axial stiffness. Thus, the bellow structure does not sustain its own weight, which implies that the actuator cannot be set along the horizontal direction (e.g., beam orientation), and it should be placed along the gravity direction. Hence, for vacuum-powered bellow actuators, both enhancing the axial stiffness, and reducing the total weight remain open challenges.

2.1.2 Pleated skin

A typical pleated airtight membrane creates chambers enabling a linear motion upon vacuum pressure, due to a similar morphological feature of the Robotic Muscle ACtuator (also called ROMAC muscle) [40], as shown in **Figure 2f** and **g**. While to the authors' knowledge no vacuum-powered PAMs exploiting this principle can be found in the literature up to date, presumably mechanical benefits would be significant rather than contractile actuators operated by only positive pressure. Indeed, the created folds distributed along the circumference of the structure allow biaxial deformations. They can be depressurized until the internal walls touch. In principle, the compressibility of the air chamber enables radial contraction, resulting in elongation of the structure upon the vacuum pressure. On the other hand, the air chambers enable the overall structure to exhibit extension upon positive pressure due to significant radial expansion. Such biaxial deformability, mainly due to the deformation in the radial direction, could play a key role in keeping the volume constant during inflation/deflation (i.e., the muscular hydrostat principle [50]), and in enlarging a linear displacement, rather than uniaxial deformation that many PAMs have accomplished so far. Moreover, the tension force that the pleated PAM can produce is adjustable. From a theoretical point of view, the tension force (F_t) generated by the imposed positive pressure (P) is strongly influenced by the number of pleats (k) [51], as follows:

$$F_t = P\frac{k}{2\pi}\sin\left(\frac{2\pi}{k}\right)l_0^2 f\left(\in, \frac{l_0}{R}\right) \tag{4}$$

where l_0 and R indicate the initial length and radius, respectively. An example of a device accomplishing this behavior is a cylindrical structure having coupled and parallel fluidic channels arranged along its circumference [20]. Due to the adjustability of radial extension and contraction, Hao et al. developed a fingerless soft gripper capable of multiple grasping modes. In order to have large deformation, such structures should have optimized pleated patterns, and they should be fabricated with the hyperelastic materials.

2.2 Origami-inspired design

Origami is a unique technique inspired by Japanese paper folding, which can be created by folding thin sheets along the predefined creases [52–54]. There are three identical crease patterns in the origami, which are the Yoshimura pattern, Miura-ori

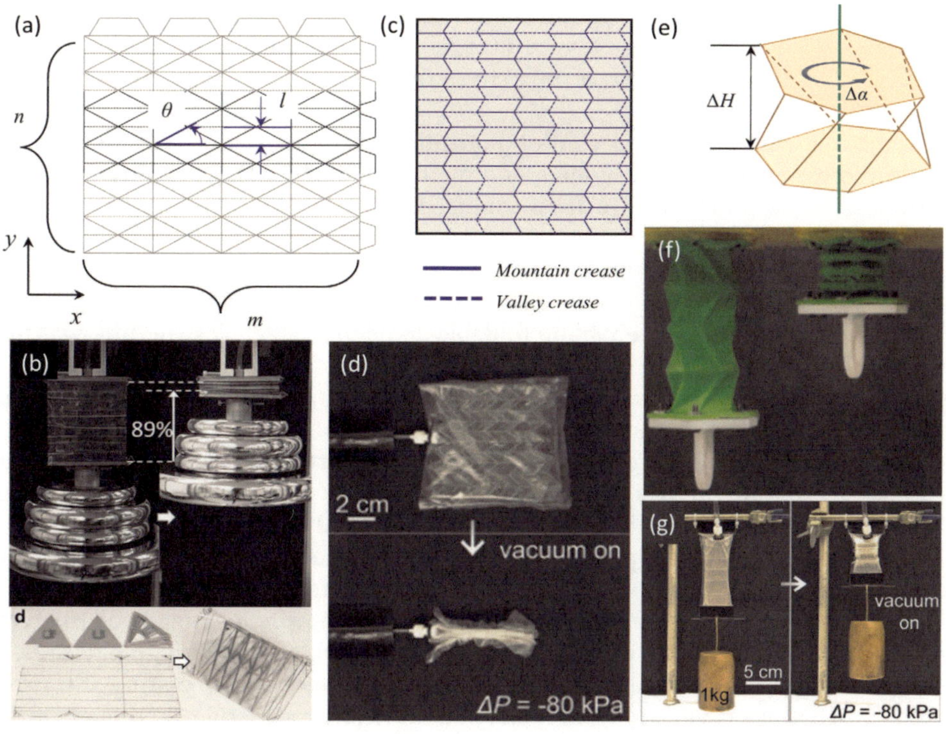

Figure 3.
Origami-inspired design approaches. (a) 2D crease pattern of the 4 x 6 Yoshimura-Ori (DoFs = 26, m = 4, n = 6). (b) an equilateral triangular shaped Origami-based vacuum PAM (OV-PAM) capable of contraction ratio up to 90% and blocking force up to 400 N. Reproduced with permission [43]. (c) a 2D crease pattern of the Miura-Ori (solid and dashed lines indicate mountain and valley creases, respectively). (d) the Miura-Ori based surface skeleton capable of a 92% contraction [57]. (e) a geometry of the crease pattern of the Kresling pattern. (f) a 3D printed origami V-PAM capable of purely axial movement due to the even number of reversed diagonal creases. Reproduced with permission [58]. (g) a lightweight origami skeleton-based actuator (2.6 g) capable of a payload of 3 kgf [57].

pattern, and Kresling (diagonal) pattern [55]. Depending on delivering motion, active or passive origami can be defined – the active origami exhibits movements through the embedded mechanism and/or structure itself. However, passive origami exhibits the motions derived from other actuation sources. Moreover, either supporting motions (i.e., structural stiffness enhancement) or achieving programmable motions can be achieved [56].

As regards the active origami, the deformability is commonly given by a function of a geometrical length (l) in the origami unit and the intersection angle of two facets (θ) (See **Figure 3a**) [59]. From the kinematics point of view, the number of creases is crucial to determine the overall structure's compliance (e.g., for the Yoshimura pattern, DOF = 9 + 4•(m-1) + (n-1), with m and n indicating the number of basic creases in x and y direction, respectively [60]). The output force can be represented [43] by

$$F = -PA_e \quad (5)$$

where A_e indicates an effective area (i.e., cross section) of the structure. Note that in the case the origami structure is made of polymers, an inflated state could be approximated as a circle due to its omnidirectional deformation. On the other hand, the deflated state needs to be rigorously modeled because the deformation occurs along the crease originated upon vacuum pressure. Therefore, to precisely predict the kinematics actuator, the effective area (i.e., cross section) should be taken into account.

In the following, an overview of origami-inspired pneumatic actuators made of different crease patterns is provided, and their design principles allowing for linear motions upon vacuum pressure are introduced.

2.2.1 Yoshimura patterned unit cell

The Yoshimura pattern, enabling foldable and deployable 3D structure, was introduced in 1955 by a Japanese scientist who first observed a buckling at thin-walled cylinders upon axial compression [53]. The Yoshimura pattern consists of identical isosceles triangles symmetrically connected in each row [56]. This way, it generates a purely translational motion element with high axial stiffness. Due to these characteristics, the Yoshimura pattern can be exploited as, both, a passive origami to reinforce movements [61], and an active origami capable of axial movement [62].

Based on the Yoshimura patterns, Martienz et al. presented novel design principles for actuators to respond to pressurization with a wide range of motions (i.e., bending, extension, contraction, twisting, etc.) [62]. A bellow morphology made of 10 creases enabled high deformability, up to 361% at 17 kPa, and the lifting of high payloads, that is, 1 kgf being 120 times its own weight (8.3 g). This method (i.e., casting of elastomer and paper coating) is versatile and it allowed fabricating an active 3D origami structure capable of movements in different directions. Moreover, to achieve both linear movements and high actuation force upon vacuum pressure, Zaghloul et al. presented Origami-inspired Semi-soft Pneumatic Actuators (OSPAs) based on the accordion and Yoshimura patterns [63]. For the OSPA with the accordion patterns, a high contraction ratio of 36% and blocking force of 124 N were achieved at a vacuum pressure of −80 kPa. Unlike the conventional Yoshimura pattern, a new design principle was introduced by modifying the Yoshimura pattern [43]. The cross section has a unique configuration as an equilateral triangle, as shown in **Figure 3b**. These morphological features allowed not only the thin film to be folded in an even and ordered manner, but also the actuator to produce a large contraction ratio of up to 99.7%, with a high blocking force of 40 kgf at a vacuum pressure of 60 kPa.

In summary, the Yoshimura patterns enable active 3D origami actuators capable of high deformability and blocking force. These mechanical features are certainly useful to achieve highly versatile and robust soft machines. Moreover, also sensory integration has been investigated. Indeed, Shen et al. embedded optical sensing solution into the origami-inspired PAM (so-called Soft Origami Optical-Sensing Actuators (SOSAs)) and presented a hybrid underwater 3-DoFs manipulator [59].

2.2.2 Miura-Ori patterned unit cell

The Miura-Ori pattern, introduced by Japanese Kogyo Miura, represents a folding technique consisting of congruent parallelograms forming a zigzag configuration in two directions [64, 65] (**Figure 3c**). The Miura folding enables high stiffness, compressibility, and extensibility [66]. Its geometry plays a key role in the mechanical properties of the folded metamaterial, exhibiting a negative Poisson's ratio between the two planar degrees of freedom [67]. Due to such extraordinary mechanical characteristics, this folding pattern was employed for the packaging and deployment of large membranes in space, such as foldable maps or solar panel deployment [68, 69]. From the architecture point of view, two different approaches are available to achieve desired linear deformations—1) a single planar Miura-Ori sheet mainly due to in-plane kinematics; or 2) a 3D folding structure (i.e., bellow) mainly due to out-of-plane kinematics.

As regards in-plane kinematics, Li et al. presented a surface (2D) skeleton made of a standard Miura-Ori pattern and demonstrated its high compressibility of up to 92% upon vacuum pressure [57], as shown in **Figure 3d**. Moreover, by exploiting asymmetrical out-of-plane motions, a 2D Miura-Ori skeleton was capable of complex motions combining both torsion and contraction. For the 3D folding structures, Reid et al. explored the bistability of the bellow pattern with the Miura-Ori folds and presented a promising technique, enabling arbitrarily complicated bellows with finely tuned fold parameters [70]. To achieve a compressible 3D structure, each bend should be paired, and thus the number of bends was even. Another example is the monothetic foldable part composed of two Miura-Ori units combined up and down. Yu et al. introduced pneumatic foldable actuators (PFAs) capable of biaxial movements upon vacuum and positive pressures [66]. This way, the compression of up to 43% was achieved at −10 kPa. However, the extensibility was limited down to 19% at 10 kPa. It is noted that the movements were mainly dependent on the length change by the folding of the actuator at low vacuum pressure, rather than the volumetric changes by positive pressure.

2.2.3 Kresling patterned unit cell

The Kresling pattern consists of a series of parallel diagonal creases defined by triangular facets [71, 72] (see **Figure 3e**). Such triangulated cylinder pattern enables bistable movements (i.e., contraction and elongation) facilitated by the buckling of the thin wall [73] while ensuring the fully foldable structure. However, this approach could suffer from mechanical failures (i.e., unbalanced deployment), resulting in coupled longitudinal and rotational motions. In principle, the work (W) required for each rotational step can be defined as a function of the elastic energy associated with axial stiffness (K_{axial}) and torsion stiffness ($K_{rotational}$). Thus, as shown in **Figure 3e**, Pagano et al. derived the overall structure stiffness given by a function of variations in rotation angle (α) and the height (H) [73], as follows:

$$K_{axial} = \frac{W^2}{\Delta H^2}, \; K_{rotational} = \frac{W^2}{\Delta \alpha^2} \tag{6}$$

Due to bistability, the actuators are generally capable of axial deformations together with coupled rotational motions *via* different core actuation sources (i.e., pneumatic [44, 74], tendon [75], magnetic [76], Shape Memory Alloy (SMA) [71], etc.). To the authors' knowledge, very few studies have demonstrated pneumatic actuators based on the Kresling pattern. Zhang et al. presented a Pneumatic and Cable-driven hybrid Linear Actuator (PCLA) capable of both thrust and tensile force [74]. The PCLA can be fully deployed (up to 200%) with a low input pressure of 2 kPa at null load and generates high force up to 70 N at 10 kPa. Herein, the diagonal creases play a crucial role in making both axial and torsional movements only for the single-story origami chamber. In this view, Vacuum-powered Soft Pneumatic Twisting Actuators (V-SPTAs) designed by adopting the diagonal creases at the single chamber could belong to the Kresling pattern class [44]. The V-SPTA was capable of high contraction force and torque. Its mechanical performance can be tunable depending on geometric parameters (i.e., initial height and initial rotation angle). For the multiple chambers, such as origami stories, the even number with reversed diagonal creases can counterbalance the rotational motions that occur at each chamber. This approach enabled the development of a 3D-printed origami V-PAM capable of pure axial movement [58], as shown in **Figure 3f**. Due to the rigid inner support rings interfaced at each module, the radial contraction can be avoided, resulting in a high compression ratio of 62% with 3 kgf payload upon vacuum pressure.

2.2.4 Origami-skeleton structure

In origami-skeleton structures, zigzag patterned structure is exploited, as a skeleton made of minimally extensible materials, such as air-tight fabrics and polyethylene [57, 77]. In principle, these compressible skeletons incorporate collapsible structures inspired by origami and/or mechanical springs. Hence, although the skin enveloping the skeletal structure has no morphological features (i.e., bellow or pleated skin), the linear movement is mainly due to the compressible skeletons. More in detail, the vacuum pressure enables the volume inside the film pouch to be evacuated, and it induces a controlled collapse in the direction guided by the skeletal structure [78]. With these findings in mind, Li et al. introduced design principles that enable linear movements and presented Fluid-driven Origami-inspired Artificial Muscles (FOAMs) by exploiting symmetrical zigzag geometry [57]. They investigated the mechanical performance of the FOAMs based on different design parameters and concluded that the maximum contraction ratio can be achieved by exploiting thin zigzag skeletons. Indeed, for the nylon fabric pouches with 0.34 mm thickness, the actuator has shown 50% linear contraction with a blocking force of 201 N. In the case of a thinner film made of polyester material (0.038 mm, with a skeleton thickness of 0.254 mm), an ultralight FOAM (2.6 g) was capable of a blocking force of up to 3 kgf, as shown in **Figure 3g**. Based on these design principles, Oguntosin et al. developed artificial muscles completely made of soft silicone rubber without any rigid parts [79]. A maximum contractile strain of 67% was achieved upon null load and a vacuum pressure of 34.5 kPa.

Another approach consists in exploiting a mechanical compression spring. Kulasekera et al. presented a Low-Profile Vacuum Actuator (LPVAc) by low-profile spring encased within a polyethylene film pouch that contracts longitudinally when vacuum pressure is applied [80]. This study highlighted that actuation was strongly dependent on the mechanical characteristics of the low-profile spring, and a high force-to-weight ratio was achieved. Indeed, the LPVAc was only 14 g, yet a high contractile strain of 65% and blocking force of 2.2 kgf (that corresponds to 160 times its own weight) were achieved. The same research group extended further this work. An airtight pouch made of a TPU (thermoplastic polyurethane) coated fabric was interfaced with a helical spring skeleton. Hence, the overall weight was reduced down to 2 g, and a high-force-weight ratio (230) was obtained [81]. More recently, they proposed a Thin-walled Vacuum Actuator (ThinVAc) and developed a multi-filament actuator, allowing simple scaling of actuation force while retaining the ability to actuator during deformation [78]. Finally, the weight of the ThinVac was only 1 g and capable of 60% linear contraction and blocking force of 5.2 N upon null load. Thus, an extremely high force-to-weight ratio of 477 was achieved.

The research on these skin-skeleton designs, demonstrating that low weight and large force-to-weight ratio are achievable, led several researchers to develop them further for wearable applications. In particular, in the rehabilitation field of infants (of six months or younger), Mendoza et al. optimized the zigzag patterned skeleton. They presented a Low-Profile Vacuum-Powered Artificial Muscle (LP-VPAM) capable of a 61% linear contraction ratio, with a lifting force of 26.4 N at a low vacuum pressure of 40 kPa [77]. Similarly, the LPVAc has shown strong potential for exploiting them in Sit-To-Stand (STS) motion assistance [80, 82].

2.3 Structural instability

Mechanical instabilities in soft PAMs and V-PAMs induce a column squirm (also called buckling), which are considered as mechanical failures in rigid body structures, and result in sudden and significant geometric changes [48, 83–85]. However, in soft robotics structural instabilities can be fruitfully exploited, and this approach shows promising results enabling new functionalities [84, 86]. Indeed, it has been reported that a reversible buckling in assemblies of elastomeric beams or films can be exploited at a variety of soft robots' applications, for example, stretchable soft electronics [87, 88], tunable metamaterials [89–91], or actuators [92]. More importantly, it is remarkable that a structure buckling allows instantaneously trigger large changes in internal pressure, deformation, shape, and exerted force [93].

In this view, it has been shown how a vacuum-driven actuator exploiting a structural instability can achieve an axial deformation due to reversible and cooperative buckling of the beams, enabling an anisotropic change in the shape of the structure [92]. More in detail, such Vacuum-Actuated Muscle-inspired Pneumatic structure (VAMP) was capable of a large longitudinal contraction (up to 40%) with a smaller horizontal deformation (5%), as shown in **Figure 4a**. Yang et al. expanded this work to develop a Shear Vacuum Actuated Machine (Shear-VAM) capable of a linear motion that works by converting the vacuum pressure, applied perpendicularly to its inextensible lateral surfaces, to a force parallel to them *via* tilted elastomeric beams [35], as shown in **Figure 4b**.

Another example is to exploit the characteristics of foams and elastomers. Roberston et al. presented a Vacuum-powered Soft Pneumatic Actuator (V-SPA) and developed 3-DoFs robotic platform [94]. A particular geometry (see **Figure 4c**) that allows inward compression and buckling of the side walls were employed to avoid possible interferences that could limit the actuator's strokes. For a single module, the V-SPA was capable of generating a blocking force of 0.92 N upon vacuum pressure of 35 kPa [34].

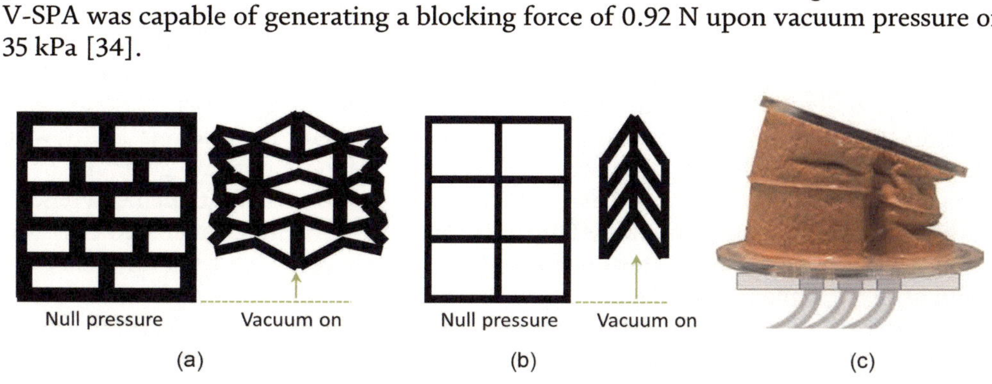

Figure 4.
(a) and (b) working principles of vacuum-powered Pneumatic Artificial Muscle (VAMP) exploiting a structural instability [92] and of Shear Vacuum Actuated Machine (Shear-VAM) [35]. (c) 3 DoFs robotic platform interfaced with vacuum-powered Soft Pneumatic Actuators (V-SPA). Reproduced with permission [94].

3. Fabrication and materials

The manufacturing process is an essential, yet delicate, step in the implementation of soft pneumatic actuators. However, from the manufacturing point of view, there are no significant differences among vacuum-powered or inflatable fluidic actuators.

Depending on the design principles, there are several approaches to fabricate V-PAMs, and an adequate fabrication protocol should ensure high reproducibility and repeatability. Due to the rapid development in additive manufacturing, modern techniques (e.g., stereolithography or multi-material 3D printing, etc.) are currently available, which enable the direct fabrication of the actuators with a good trade-off between material consumption and cost-effectiveness. Moreover, such additive manufacturing techniques are also widely used to fabricate molds and active elements typical of the hybrid manufacturing approach.

In this section, the fabrication approaches that are commonly used for implementing V-PAMs are overviewed (with mention to the employed materials), including molding manufacturing, additive manufacturing, and hybrid techniques. Their advantages and limitations are discussed and compared.

3.1 Molding manufacturing

Molding manufacturing is defined as the process of shaping a liquid or a raw material using a rigid container called mold or matrix, which has a negative geometry of the desired shape. Due to the advent of 3D printing, laser cutting, and CNC milling, fabricating the mold has become affordable and relatively fast. Here, depending on the complexity of the object, different molding approaches need to be considered (i.e., casting, blow molding, injection molding, etc.), while only a few of them have been widely addressed to produce soft actuators. In literature, it is frequent to find examples of inflatable fluidic actuators [95–98] and vacuum-powered actuators [44, 92, 99] mainly built *via* casting or injection molding. Generally, in both approaches, due to the shape complexity of vacuum actuators, articulated molds, and multiple steps are required. To better understand these molding techniques, and to describe some relative case studies for V-PAMs, three main phases in fabrication need to be considered—mold design and fabrication, molding, demolding, and assembly.

A correct design and construction of the mold is the first step to succeed in this process. A good practice is to design the mold to be reusable while simplifying the demolding phase and avoiding material waste due to sacrificial molds. This way the risk of damaging the samples during extraction is minimized. To this aim, materials used for the mold should be investigated, taking into account their compatibility with casting materials, and curing parameters. Given that soft actuators are generally built from silicon-rubber or urethane-based silicon, then Acrylonitrile butadiene styrene (ABS), PolyTetraFluoroEthylene (PTFE), PolyAmide (PA), and Chlorinated PolyEthylene (CPE) are good candidates for the molds. Alternatively, PolyLActide (PLA) could be used, while curing elastomers can be completed only at room temperature due to its thermomechanical characteristics (e.g., glass transition temperature T_g).

The second step is to pour the elastomers into the mold and to determine an adequate curing process with specific temperatures and timing. As shown in **Figure 5a**, based on the mold filling methodology, either casting (where the liquid material is driven by gravity in the mold cavities), or injection molding (where the material is forced into the mold by pressurization) can be used. Once the mold is filled, it must be placed in a vacuum chamber to remove air bubbles that would instead remain trapped in the cured material affecting the final mechanical characteristics of the same.

The last step is to extract the cured parts without any damage. Where proper mold design is insufficient to ensure a correct extraction, synthetic lubricants (e.g., Teflon coating) can help to avoid failures (e.g., delamination at boned layers). Finally, once all components are ready, the actuators are assembled in the desired configuration by using uncured materials or silicone adhesives.

With the aforementioned procedures in molding fabrication, Balak et al. developed a vacuum-powered omnidirectional soft robotic actuator (OSRA) [99]. A two-step molding process was employed. The core structure and cavities of the OSRA were first obtained by using three mold parts. Once the first part was cured, the actuator was completed by putting the cured part into a second mold to make the base seal. Similarly, in ref. [44], the authors employed the molding fabrication technique and could obtain a vacuum soft pneumatic twisting actuator (V-SPTA). First, the body and bottom parts of the V-SPTA were made by casting and, in a second step, attached through a silicone adhesive layer. The same two-step procedure was applied to

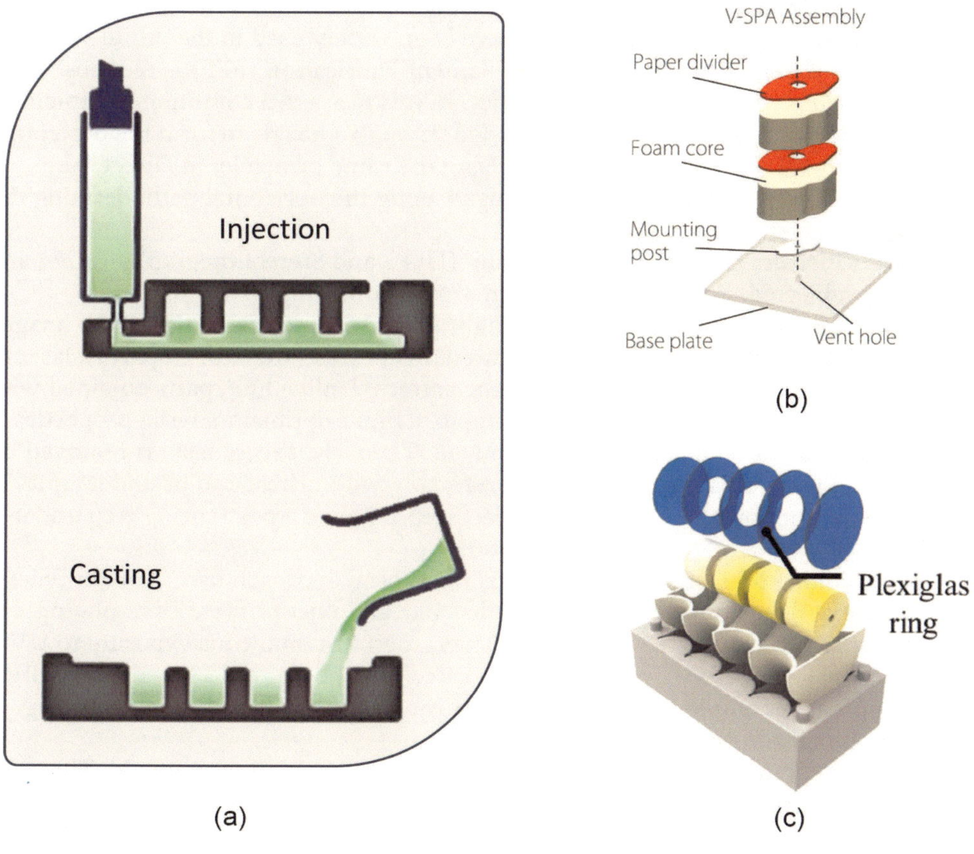

Figure 5.
Different fabrication techniques. (a) shows different filling methodologies - Injection and casting molding, respectively. The hybrid fabrication processes combining two or more techniques - (b) encasing the foam core with the elastomer and smoothly merging each module by interfacing paper dividers. Reproduced with permission [94]. (c) a bellow type actuator (UH-PAM [38]) with foam modules and Plexiglass© rigid rings interfaced at each convolution of the elastomeric bellow skin [38].

fabricate the buckling actuators shown in refs. [44, 58, 79, 92, 99]. However, the silicone curing was performed in a controlled environment at 60°C for 10 minutes, reducing the fabrication time. Finally, Oguntosin and Akindele [79] show a remarkable example of a soft vacuum actuator fabricated entirely from silicon rubber. The approach followed the molding method, but unlike the previous examples [44, 79, 99], two different elastomeric materials were smoothly interfaced in one final structure.

3.2 Additive manufacturing

Additive manufacturing (AM), also known as three-dimensional (3D) printing, encompasses several emerging techniques based on the same principle: starting from a 3D model, replicate the structure by sequentially adding layers of material until model completeness. In literature, different fabrication protocols have been investigated to obtain soft actuators operated by either positive [100–104] or negative pressure [37, 105, 106].

Table A2 (in the Appendix) summarizes different AM processes defined by ISO/ ASTM 52900:2021. Of these, only Vat PhotoPolymerization (VPP), Material EXtrusion (MEX), and Material Jetting (MJT) have been widely used in the production of soft actuators [107]. In particular, Fused Filament Fabrication (FFF) is the most widely used MEX technology in soft robotics. In this process, a continuous filament of thermoplastic material is melted and extruded through a nozzle to form the horizontal layers of the 3D structure [108, 109]. Using the same principle, in Direct Ink Writing (DIW), a liquid-phase ink is arranged along the horizontal path, defining the 3D structures layer by layer [110].

On the VPP side, Digital Light Processing (DLP) and Stereolithography (SLA) are commonly subjected. Both use photopolymer resin, differing curing methods [109, 111]. The DLP printer uses a digital light projector screen that flashes an image of the layer that defines the surface to be cured. In contrast, the SLA employs a laser to draw and cure the layer following the surface pattern. Unlike FFF, parts obtained with DLP and SLA need adequate post-processing to achieve optimal material properties. First, all components should be gently removed. Then, the excess resin is removed by using ethanol (bio) or isopropyl alcohol, preferably with ultrasound or under agitation. Once the supports are removed, the last step requires a post-curing step under high-power UV for a few minutes at about 65°C.

Finally, among MJT processes, PolyJet is the typical approach used. PolyJet relies on a sequential deposition of material droplets like an inkjet printer. First, photopolymer droplets are accumulated to form the layer, then instantaneous exposure to UV light cures the layer before a new one is deposited [112]. Stano et al. recently classified the AD approaches employed to fabricate soft robots in rapid mold manufacturing, hybrid, and total additive manufacturing [107].

In rapid mold manufacturing, the AM limits the production of the mold. The material used in this process does not define the properties of the actuator, which instead are linked to the casted material in a second phase, usually silicone or elastomer. The FFF is the standard technique used in this approach, and ABS and PLA filaments are the typical materials used. The resolution of the 3D printer plays an essential role in the presence of small details. Hence, even though there can be inherent shortcomings (e.g., the high cost of the printing system, the post-processing steps), professional 3D printers are preferred due to the higher resolution along the z-axis. Indeed, DLP or SLA techniques have been used, to implement complex structures that need to be fabricated with high precision.

Furthermore, the hybrid approach combines AM technologies with traditional manufacturing approaches. Unlike in molding, the elements printed with this approach are embedded in the final object and influence the kinematics of the actuators.

Total Additive Manufacturing (TAM) has attracted much interest in fabricating soft actuators. In this approach, only AD techniques are exploited. Depending on the actuator's complexity, the TAM could follow either a modular or monolithic approach. For the former, all "modules" are printed separately and assembled into the final configuration only in a second step, whereas the latter involves TAM to obtain the actuator by a single printing cycle. Combined with multi-material 3D printing, monolithic TAM allows printing actuators made of two or more materials in a single cycle. The same approach permits embedding sensing elements in actuators, enhancing the mechanical synergy between sensing and movements. In parallel to selecting adequate printing methods among MEX, VPP, and MJT, the material properties need to be considered to accomplish desired mechanical performance. Otherwise, custom-made printers could be developed. Indeed, Byrne et al. proposed a custom setup to exploit the advantages of multi-material 3D printing [113]. Combining fused deposition modeling and a paste extrusion printer, they fabricated soft actuators capable of contracting, bending, or twisting in a single printing run. Similarly, by exploiting a commercial multi-material 3D printer (Ultimaker S3), Stano et al. presented a monolithic bending PneuNets made of TPU 80A and TPU 95A [114].

More recently, the TAM approach has been successfully applied to fabricate vacuum-driven actuators by using fused filament fabrication. In ref. [106], Tawk et al. proposed a bioinspired 3D printable Soft Vacuum Actuators (SOVAs). They showed the versatility functionalities of the AM fabrication by demonstrating various robotic applications (i.e., locomotion robots, grippers, and artificial muscles), by applying both monolithic and modular approaches. Similarly, they presented a Linear Soft Vacuum Actuator (LSOVA) fabricated in a single step using an open-source 3D printer and a commercial TPU filament NinjaFlex (NinjaTek, USA) [37], as shown in **Figure 2d**. While SOVA and L-SOVA are bellow-like and linear vacuum actuators, respectively, the solution proposed by Zhang et al. in [58] is a fully 3-D printed origami-inspired VPAM. Same as molding manufacturing, this demonstrates the versatility of the AM approach, which makes it suitable for different designs.

In summary, the success rate in fabrications exploiting AM is mainly dependent on determining adequate printing parameters. Indeed, an error that could occur at the beginning phase of printing significantly affects the entire process, resulting in printing failures. In particular, for FFF, the bed temperature should be identified to ensure proper layer adhesion without overheating the first layer. At the same time, the size of the first layer should be chosen to accomplish the best quality and airtightness. Avoiding high or low retraction values may help prevent under-extrusion or printed residue. Extrusion temperature, printing speed, and cooling speed deeply affect layer bonding, thus the quality and integrity of the printed part. Finally, an infill density related to the overall structure stiffness enables determining the strength of actuators.

3.3 Hybrid and alternative manufacturing

Hybrid manufacturing combines two or more fabrication techniques. Typically, a tradeoff between 3D printing and other fabrication processes, such as molding manufacturing, is addressed. Qi et al. encapsulated 3D-printed elements made of custom material into a silicone structure obtained through mold fabrication [115]. Similarly, in ref. [116], a soft actuator with stiffness and shape modulation was obtained by embedding a 3D-printed Conductive PolyLactic Acid (CPLA) layer in a soft pneumatic actuator made of silicone. Robertson et al. in refs. [94, 117] fabricated a V-PAMs without employing molding and additive manufacturing, as shown in **Figure 5b**. First, by using a laser cutting machine, the foam chambers and rigid layers were cut. Then, by gluing them, the core structure of the actuator was assembled, manually coated, and sealed by using Elastosil M4601. **Figure 5c** shows the hybrid fabrication process proposed by the authors in order to develop the UH-PAM [48] (described in Section 2.1.1). Unlike previous examples, after patterning by laser cutting, the open-cell foams (Polyurethane) were integrated with Plexiglass©rigid rings, and with a silicone (Dragon Skin™ 30) skin having a bellow structure, previously molded by casting. Another example belongs to a completely hybrid concept compromising rigid and soft materials [78, 81]. In the V-PAM fabrication, a helical spring core was combined with a pouch, manually assembled from polyethylene (PE) film or TPU fabric. These studies show that hybrid manufacturing has a strong potential, to break down the barrier between rigid and soft materials and to allow them interfacing.

Despite the promising mechanical performance that the hybrid approach delivers to the soft actuators, it suffers from a decreased process controllability since a higher number of separate, and usually manual, fabrication steps are needed, resulting in reduced repeatability and reproducibility. For example, in refs. [78, 81], the repeatability, thus the homogeneity of two different actuators, cannot be ensured since both depend on the manual skills of the operator. Similarly, in ref. [94, 117] using a manual deposition of the core, the homogeneity of the silicone layer could be unevenly distributed.

3.4 Materials

In addition to exploring fabrication protocols, selecting adequate soft materials is a key enabler for soft robot bodies. Although the meaning of the term "soft" has been contentious, in literature soft robots are defined as systems composed of materials with Young's modulus ranging from 10 kPa to 1 GPa, similar to soft biological materials (i.e., skin or muscle tissue) [118]. In this view, V-PAMs can consist of a wide range of materials (i.e., polymer, rubber-like silicone) that exhibit hyperelastic behavior with a large deformation ratio, or inextensible thin layered materials (i.e., paper, fabric) that could belong to both soft and rigid materials. In this section, some highlights to address an adequate selection of the materials are provided, referring to the main material candidates that are interesting for the fabrication of V-PAMs.

Given that soft materials strongly influence the actuator mechanical properties as well as its kinematic behavior, selecting adequate materials is mainly driven by the predicted deformations (and stresses) that are required along the overall structure. To this aim, for the linear motion, implicit (i.e., Finite element method (FEM)) or explicit numerical approaches (e.g., Euler beam theorem based) were employed to predict desired kinematic trajectories versus input vacuum pressure [119–122]. Moreover, for a V-PAM to achieve high compliance, in addition to the skin material intrinsic compliance, air impermeability is necessary. From the material point of view, elastomeric polymers or silicones can withstand large deformations, in some cases like EcoFlex™ 00-30 up to 900%, without undergoing permanent plastic deformation. Moreover, both air impermeability and compliance can be obtained by exploiting compositions and depositions [123, 124].

As representative solutions, silicones and urethane-based rubbers can be used to realize V-PAMs where molding fabrication is employed. For those hyperelastic constitutive materials, two components (i.e., elastomers, catalyst, or hardener) are mixed at precise mixing ratios (typically 1:1) before degassing and curing. From the mechanical point of view, different elongation at break and Young's modulus can be achievable, depending on materials. For example, platinum-catalyzed silicones (EcoFlex™, DragonSkin™, etc.) are commercially available and exhibit elongation at break ranging from 800% to 1,000% and Young's modulus ranging from few hundreds kPa to 50 MPa; whereas the mechanical characteristics of Polydimethylsiloxane (PDMS, Sylgard184) can be tuned by using different elastomer/cross-linking agent ratio (e.g., 10:1, 20:1, 30:1).

As regards additive manufacturing, filaments of TPU, such as NinjaFlex® TPU or Ultimaker TPU, are the most used to fabricate soft actuators (including both inflatable fluidic actuators and V-PAMs) by the TAM approach. In general, TPU filaments are characterized by hardness in the 75–90 Shore A range, with an elongation at the break that could reach almost 700%. Alternatively, the Thermoplastic Polyether-Polyurethane elastomer, such as the Recreus FilaFlex, shows a similar characteristic, offering elongation at break up to 950% and hardness in the range 60–95 shore A.

Table 1 provides a brief overview of some of the most common materials used in the fabrication of V-PAMs. In terms of elongation at break, both silicone and filaments offer good performance, showing values that can range from 100% up to a maximum of 1000%. On the contrary, from the point of view of Young's modulus, the V-PAM obtained with silicone shows values 10 times lower than the counterpart obtained with the TAM approach. This means that for the same structure, actuators obtained by casting are always softer than those obtained by 3D printing, which resulting more rigid require, an higher actuation power to achieve the same behavior. On the other hand, higher stiffness can lead to solutions that can support a higher payload. For this reason, the choice of material is a compromise between performance, feasibility of the concept and operating conditions. The latter point in particular makes it difficult to define a priori which material is best based on the design used alone, making careful analysis necessary in the design phase.

4. Conclusion and outlook

This review introduces the promising potential of V-PAMs for implementing complex soft machines achieving robustness and fail-safe operation. In contrast to the inflatable fluidic actuators, given that vacuum pressure induces a volumetric contraction, the linear deformation is strongly influenced by the contraction in the radial direction. Thus, their structures and geometries can avoid stiffening and/or an increase in stress during movement, at the same time ensuring a good level of compactness and robustness. In particular, due to their reliability [37, 92, 106], the V-PAMs proposed to show a lifespan that varies from a minimum of 21,500 cycles of LSOVAs up to more than 1,000,000 cycles of VAMPs. Furthermore, being based on volumetric contraction, V-PAMs results suitable for operating in both large and limited spaces, oppositely to PAMs, which exploiting a volumetric expansion cannot properly work in confined environments.

Nevertheless, there are still several open challenges to be addressed. Indeed, new trade-offs among design principles and desired mechanical performance need to be found to enable energy efficiency in actuators with high power-to-weight ratio. Also, scalability remains barely impossible, mainly due to the square-cube law.

Given that upon vacuum pressure, the overall structure stiffness is mainly dependent on its bulk state, determining adequate materials is a key aspect to achieve the desired actuation force. Thus, exploiting different materials (multi-material) approaches that allow for enhancing the axial stiffness at a global level have been widely addressed [125–128] by smoothly interfacing them with the structure. However, this approach could lead to an increase in the overall structure weight, as well as material failures (i.e., delamination). Thereby, there is the quest for new materials and/or investigation of novel design principles.

An interesting approach is to employ foam (like in the UH-PAM [38]) as a constituent material, or porous structures where the air-cells dimensions can be decided upon the desired stiffness. Notably, given the recent advances in foam-like sensing,

	Young's module [MPa]	**Shore hardness [A]**	**Elongation at break [%]**	**Manufacturing**		
				Molding	**Additive**	**Hybrid**
Silicones						
Ecoflex ™ 00-10	0.55	10	800	•		•
Ecoflex ™ 00-20	0.55	20	845	•		•
Ecoflex ™ 00-30	0.69	30	900	•		•
DragonSkin™ 10,	0.151	10	1,000	•		•
DragonSkin™ 20,	0.338	20	620	•		•
DragonSkin™ 30,	0.593	30	364	•		•
Sylgard 184 (10:1)	1.32	43	100	•		•
Elastosil® M4601	0.262	28	700	•		•
Filament (TAM)						
Ultimaker TPU	26	95	580		•	•
NinjaFlex© TPU	12	85	660		•	•
Chinchilla™ TPE	34	75	600		•	•
FilaFlex 60A (TPU)	26	63	950		•	•
FilaFlex 82A (TPU)	22	82	650		•	•
FilaFlex 70A (TPU)	32	70	900		•	•
	Young's module [GPa]	**Shore hardness [D]**	**T_g [°C]**	**Manufacturing**		
				Molding	**Additive**	**Hybrid**
Filament (molds)						
Ultimaker ABS	1.618	76	90–105		•	•
Ultimaker CPE+	1.128	75	90–110		•	•
Ultimaker PLA	2.346	83	55–65		•	•
PolyLite™ ABS	2.174	—	101		•	•
PolyLite™ PLA	2.636	—	63		•	•

Table 1.
Materials' properties and the compatibility with different V-PAMs manufacturing.

(e.g., capacitive [129], inductive sensing [130], resistive sensing [131, 132], etc.), it can be construed that, in a near future, new designs should consider such transduction mechanisms embedded in V-PAMs. This approach seems promising to develop soft machines with embodied intelligence *via* proprioceptive and/or exteroceptive sensory feedback [133].

Indeed, due to the rapid development in additive manufacturing, designing and fabricating 3D architectures made of metamaterials has been demonstrated [134–137]. Hence, it can be expected for them to play a key role in the future development of soft actuation, allowing the soft machines to exhibit desired kinematic performance while reducing overall weight due to the tessellated topology and morphology. For this to happen, the quest is for materials that possess high extensibility (>500%), Young's modulus in the range of tens of MPa, and high reliability and low hysteresis.

In conclusion, in any case, it will be important to address designs that exploit both positive and negative pressures, rather than choosing one of them. This way motions in different directions could be achieved with the same actuator, for example, elongation, contraction, bending, twisting, also inspiring from natural models. For example, this is the case of developing innovative continuum manipulators inspired from elephant trunks, like in the EU project PROBOSCIS [138], which aim is to develop "soft" yet "strong" perceptive soft machines.

Acknowledgements

This work has received funding from the European Union's Horizon 2020 research and innovation program under grant agreement No. 863212 (PROBOSCIS project).

Conflict of interest

The authors declare no conflict of interest.

Appendices and nomenclature

Nomenclatures	Descriptions	Nomenclatures	Descriptions
D	Outer diameter	TPU	Thermoplastic polyurethane
d	Inner diameter of the bellow	PDMS	Polydimethylsiloxane
D_m	Mean diameter of the bellow	PE	Polyethene
w	Convolution depth	AM	Additive Manufacturing
E_b	Young's modulus of the material used for the bellow	VPP	Vat Polymerization
t_p	Thickness of skin	MEX	Material Extrusion
μ	Poisson's ratio	MJT	Material Jetting
n	Number of bellow plies (in case of multiple plies)	DLP	Digital Light Processing
N	Number of convolutions	SLA	Stereolithography

Nomenclatures	Descriptions	Nomenclatures	Descriptions
ABS	Acrylonitrile butadiene styrene	FFF	Fused Filament Fabrication
PTFE	Polytetrafluoroethylene	DIW	Direct Ink Writing
PA	Polyamide	TAM	Total additive manufacturing
CPE	Chlorinated polyethylene	T_g	Glass transition temperature
PLA	Polylactide		

Table A1.
Nomenclatures.

Process name	Operative principle
Binder Jetting (BJT)	A liquid bonding agent is selectively deposited to join powder materials
Direct Energy Deposition (DED)	Focused thermal energy is used to fuse materials by melting as they are being deposited
Material Extrusion (MEX)	Material is selectively dispensed through a nozzle or orifice
Material Jetting (MJT)	Droplets of feedstock material are selectively deposited
Powder Bed Fusion (PBF)	Thermal energy selectively fuses regions of a powder bed
Sheet Lamination (SHL)	Sheets of material are bonded to form a part
Vat Photopolymerization (VPP)	Liquid photopolymer in a vat is selectively cured by light-activated polymerization
Binder Jetting (BJT)	A liquid bonding agent is selectively deposited to join powder materials
Direct Energy Deposition (DED)	Focused thermal energy is used to fuse materials by melting as they are being deposited
Material Extrusion (MEX)	Material is selectively dispensed through a nozzle or orifice
Material Jetting (MJT)	Droplets of feedstock material are selectively deposited
Powder Bed Fusion (PBF)	Thermal energy selectively fuses regions of a powder bed
Sheet Lamination (SHL)	Sheets of material are bonded to form a part
Vat Photopolymerization (VPP)	Liquid photopolymer in a vat is selectively cured by light-activated polymerization

Table A2.
Different printing processes and their operation principles.

Author details

Seonggun Joe, Federico Bernabei and Lucia Beccai*
Istituto Italiano di Tecnologia, Genova, Italy

*Address all correspondence to: Lucia.beccai@iit.it

References

[1] Xu W, Chen J, Lau HY, Ren H. Data-driven methods towards learning the highly nonlinear inverse kinematics of tendon-driven surgical manipulators. The International Journal of Medical Robotics and Computer Assisted Surgery. 2017;**13**(3):e1774

[2] Kastor N, Mukherjee R, Cohen E, Vikas V, Trimmer BA, White RD. Design and manufacturing of tendon-driven soft foam robots. Robotica. 2020;**38**(1): 88-105

[3] Pabst O, Hölzer S, Beckert E, Perelaer J, Schubert US, Eberhardt R, et al. Inkjet printed micropump actuator based on piezoelectric polymers: Device performance and morphology studies. Organic Electronics. 2014;**15**(11): 3306-3315

[4] Kofod G, Wirges W, Paajanen M, Bauer S. Energy minimization for self-organized structure formation and actuation. Applied Physics Letters. 2007; **90**(8):081916

[5] Ji X, Liu X, Cacucciolo V, Imboden M, Civet Y, El Haitami A, et al. An autonomous untethered fast soft robotic insect driven by low-voltage dielectric elastomer actuators. Science. Robotics. 2019;**4**(37):eaaz6451

[6] Akbari S, Sakhaei AH, Panjwani S, Kowsari K, Serjouei A, Ge Q. Multimaterial 3D printed soft actuators powered by shape memory alloy wires. Sensors and Actuators A: Physical. 2019; **290**:177-189

[7] Huang X, Kumar K, Jawed MK, Mohammadi Nasab A, Ye Z, Shan W, et al. Highly dynamic shape memory alloy actuator for fast moving soft robots. Advanced Materials Technologies. 2019;**4**(4):1800540

[8] Lendlein A. Fabrication of reprogrammable shape-memory polymer actuators for robotics. Science. Robotics. 2018;**3**(18):eaat9090

[9] Ma S, Zhang Y, Liang Y, Ren L, Tian W, Ren L. High-performance ionic-polymer–metal composite: Toward large-deformation fast-response artificial muscles. Advanced Functional Materials. 2020;**30**(7):1908508

[10] Festo. TentacleGripper 2022. Available from: https://www.festo.com/us/en/e/about-festo/research-and-development/bionic-learning-network/highlights-from-2015-to-2017/tentaclegripper-id_33321/?siteUid=fox_us&siteName=Festo+USA

[11] Robotics S. mGrip 2022. Available from: https://www.softroboticsinc.com/products/mgrip-modular-gripping-solution-for-food-automation/

[12] Polygerinos P, Wang Z, Galloway KC, Wood RJ, Walsh CJ. Soft robotic glove for combined assistance and at-home rehabilitation. Robotics and Autonomous Systems. 2015;**73**: 135-143

[13] Nguyen PH, Zhang W. Design and computational modeling of fabric soft pneumatic actuators for wearable assistive devices. Scientific Reports. 2020;**10**(1):1-13

[14] Maeder-York P, Clites T, Boggs E, Neff R, Polygerinos P, Holland D, et al. Biologically inspired soft robot for thumb rehabilitation. Journal of Medical Devices. 2014;**8**(2):020933

[15] Manfredi L, Capoccia E, Ciuti G, Cuschieri A. A soft Pneumatic Inchworm Double balloon (SPID) for colonoscopy. Scientific Reports. 2019;**9**(1):1-9

[16] Okayasu H, Okamoto J, Fujie MG, Umezu M, Iseki H. Development of a hydraulic-driven flexible manipulator for neurosurgery. In: International Congress Series. 2003

[17] Watanabe Y, Maeda M, Yaji N, Nakamura R, Iseki H, Yamato M, et al. Small, soft, and safe microactuator for retinal pigment epithelium transplantation. In: 2007 IEEE 20th International Conference on Micro Electro Mechanical Systems (MEMS). 2007

[18] Chen C-Y, May KP, Yeow C-H. 3D printed soft extension actuator. In: IEEE 4th International Conference on Soft Robotics (RoboSoft). 2021

[19] Shintake J, Rosset S, Schubert B, Floreano D, Shea H. Versatile soft grippers with intrinsic electroadhesion based on multifunctional polymer actuators. Advanced Materials. 2016; **28**(2):231-238

[20] Hao Y, Biswas S, Hawkes EW, Wang T, Zhu M, Wen L, et al. A multimodal, enveloping soft gripper: Shape conformation, bioinspired adhesion, and expansion-driven suction. IEEE Transactions on Robotics. 2020; **37**(2):350-362

[21] Zongxing L, Wanxin L, Liping Z. Research development of soft manipulator: A review. Advances in Mechanical Engineering. 2020;**12**(8): 1687814020950094

[22] Kastor N, Vikas V, Cohen E, White RD. A definition of soft materials for use in the design of robots. Soft Robotics. 2017;**4**(3):181-182

[23] Gollob SD, Park C, Koo BHB, Roche ET. A modular geometrical framework for modelling the force-contraction profile of vacuum-powered soft actuators. Frontiers in Robotics and AI. 2021;**8**:606938

[24] Veale AJ, Xie SQ, Anderson IA. Characterizing the Peano fluidic muscle and the effects of its geometry properties on its behavior. Smart Materials and Structures. 2016;**25**(6):065013

[25] Wang L, Wang Z. Mechanoreception for soft robots via intuitive body cues. Soft Robotics. 2020;**7**(2):198-217

[26] Digumarti KM, Conn AT, Rossiter J. Euglenoid-inspired giant shape change for highly deformable soft robots. IEEE Robotics and Automation Letters. 2017; **2**(4):2302-2307

[27] Goulbourne N, Son S, Fox J. Self-sensing McKibben actuators using dielectric elastomer sensors. In: Electroactive Polymer Actuators and Devices (EAPAD). 2007

[28] Tondu B, Lopez P. Modeling and control of McKibben artificial muscle robot actuators. IEEE Control Systems Magazine. 2000;**20**(2):15-38

[29] Chou C-P, Hannaford B. Measurement and modeling of McKibben pneumatic artificial muscles. IEEE Transactions on Robotics and Automation. 1996;**12**(1): 90-102

[30] Al-Ibadi A, Nefti-Meziani S, Davis S. The design, kinematics and torque analysis of the self-bending soft contraction actuator. Actuators; 2020

[31] Daerden F, Lefeber D. The concept and design of pleated pneumatic artificial muscles. International Journal of Fluid Power. 2001;**2**(3):41-50

[32] Verrelst B. A dynamic walking biped actuated by pleated pneumatic artificial muscles: Basic concepts and control

issues. PhD Thesis Vrije Universiteit Brussel. 2005

[33] Daerden F, Lefeber D. Pneumatic artificial muscles: Actuators for robotics and automation. European Journal of Mechanical and Environmental Engineering. 2002;**47**(1):11-21

[34] Agarwal G, Robertson MA, Sonar H, Paik J. Design and computational modeling of a modular, compliant robotic assembly for human lumbar unit and spinal cord assistance. Scientific Reports. 2017;**7**(1):1-11

[35] Yang D, Verma MS, Lossner E, Stothers D, Whitesides GM. Negative-pressure soft linear actuator with a mechanical advantage. Advanced Materials Technologies. 2017;**2**(1):1600164

[36] Mott PH, Dorgan JR, Roland C. The bulk modulus and Poisson's ratio of "incompressible" materials. Journal of Sound and Vibration. 2008;**312**(4–5): 572-575

[37] Tawk C, Spinks GM, in het Panhuis M, Alici G. 3D printable linear soft vacuum actuators: Their modeling, performance quantification and application in soft robotic systems. IEEE/ASME Transactions on Mechatronics. 2019;**24**(5):2118-2129

[38] Joe S, Totaro M, Wang H, Beccai L. Development of the ultralight hybrid pneumatic artificial muscle: Modelling and optimization. PloS One. 2021;**16**(4): e0250325

[39] Daerden F. Conception and Realization of Pleated Pneumatic Artificial Muscles and Their Use as Compliant Actuation Elements. Belgium: Vrije Universiteit Brussel; 1999

[40] Immega G, Kukolj M. Axially contractable actuator. Google Patents; 1990

[41] Snedden N. Analysis and design guidance for the lateral stiffness of bellows expansion joints. Thin-walled Structures. 1985;**3**(2):145-162

[42] Becht C IV. An evaluation of EJMA stress calculations for unreinforced bellows. Journal of Pressure Vessel Technology. 2002;**124**(1):124-129

[43] Lee J-G, Rodrigue H. Origami-based vacuum pneumatic artificial muscles with large contraction ratios. Soft Robotics. 2019;**6**(1):109-117

[44] Jiao Z, Ji C, Zou J, Yang H, Pan M. Vacuum-powered soft pneumatic twisting actuators to empower new capabilities for soft robots. Advanced Materials Technologies. 2019;**4**(1): 1800429

[45] Yang HD, Greczek BT, Asbeck AT. Modeling and analysis of a high-displacement pneumatic artificial muscle with integrated sensing. Frontiers in Robotics and AI. 2019;**5**:136

[46] Hashem R, Stommel M, Cheng L, Xu W. Design and characterization of a bellows-driven soft pneumatic actuator. In: IEEE/ASME Transactions on Mechatronics. 2020

[47] Felt W, Robertson MA, Paik J. Modeling vacuum bellows soft pneumatic actuators with optimal mechanical performance. In: IEEE International Conference on Soft Robotics (RoboSoft). 2018

[48] Joe S, Wang H, Totaro M, Beccai L. Development of ultralight hybrid pneumatic artificial muscle for large contraction and high payload. In: 3rd IEEE International Conference on Soft Robotics (RoboSoft). 2020

[49] Belforte G, Eula G, Ivanov A, Visan AL. Bellows textile muscle. The

Journal of The Textile Institute. 2014; **105**(3):356-364

[50] Smith KK, Kier WM. Trunks, tongues, and tentacles: Moving with skeletons of muscle. American Scientist. 1989;**77**(1):28-35

[51] Casi DV. Development of the production process of PPAM: Ph. D. Thesis, Universidad Publica de Navarra; 2009

[52] Rafsanjani A, Bertoldi K, Studart AR. Programming soft robots with flexible mechanical metamaterials. arXiv preprint arXiv:190600306. 2019

[53] Yoshimura Y. On the mechanism of buckling of a circular cylindrical shell under axial compression. 1955

[54] Chen Y, Peng R, You Z. Origami of thick panels. Science. 2015;**349**(6246): 396-400

[55] Paez L, Granados M, Melo K, Conceptual design of a modular snake origami robot. 2013 IEEE International Symposium on Safety, Security, and Rescue Robotics (SSRR); 2013

[56] Onal CD, Wood RJ, Rus D, Towards printable robotics: Origami-inspired planar fabrication of three-dimensional mechanisms. 2011 IEEE international conference on robotics and automation; 2011

[57] Li S, Vogt DM, Rus D, Wood RJ. Fluid-driven origami-inspired artificial muscles. Proceedings of the National Academy of Sciences. 2017;**114**(50): 13132-13137

[58] Zhang Z, Fan W, Chen G, Luo J, Lu Q, Wang H, A 3D printable origami vacuum pneumatic artificial muscle with fast and powerful motion. 2021 IEEE 4th International Conference on Soft Robotics (RoboSoft); 2021

[59] Shen Z, Zhao Y, Zhong H, Tang K, Chen Y, Xiao Y, et al. Soft origami optical-sensing actuator for underwater manipulation. Frontiers in Robotics and AI. 2021;**7**:219

[60] Woerd JD, Chudoba R, Hegger J, Single-curved shell structure made out of textile-reinforced concrete plate using a folding technique. Proceedings of IASS Annual Symposia; 2013

[61] Paez L, Agarwal G, Paik J. Design and analysis of a soft pneumatic actuator with origami shell reinforcement. Soft Robotics. 2016;**3**(3):109-119

[62] Martinez RV, Fish CR, Chen X, Whitesides GM. Elastomeric origami: Programmable paper-elastomer composites as pneumatic actuators. Advanced Functional Materials. 2012; **22**(7):1376-1384

[63] Zaghloul A, Bone GM. 3D shrinking for rapid fabrication of origami-inspired semi-soft pneumatic actuators. IEEE Access. 2020;**8**:191330-191340

[64] Miura K. Method of packaging and deployment of large membranes in space. The Institute of Space and Astronautical Science Report. 1985;**618**: 1-9

[65] Stavric M, Wiltsche A. Investigations on quadrilateral patterns for rigid folding structures–folding strategies-rigid and curved folding. 2013

[66] Yu M, Yang W, Yu Y, Cheng X, Jiao Z, A crawling soft robot driven by pneumatic foldable actuators based on Miura-ori. Actuators; 2020

[67] Schenk M, Guest SD. Geometry of Miura-folded metamaterials.

Proceedings of the National Academy of Sciences. 2013;**110**(9):3276-3281

[68] Horner G, Elliott M, A fabrication and deployment approach for a Miura-ori solar sail model. 43rd AIAA/ASME/ASCE/AHS/ASC Structures, Structural Dynamics, and Materials Conference; 2002

[69] Horner G, Wright T, Laue G, Miura-Ori solar sail packaging concept development and deployment demonstration. 39th AIAA/ASME/SAE/ASEE Joint Propulsion Conference and Exhibit; 2003

[70] Reid A, Lechenault F, Rica S, Adda-Bedia M. Geometry and design of origami bellows with tunable response. Physical Review E. 2017;**95**(1):013002

[71] Nayakanti N, Tawfick SH, Hart AJ. Twist-coupled kirigami cells and mechanisms. Extreme Mechanics Letters. 2018;**21**:17-24

[72] Jianguo C, Xiaowei D, Ya Z, Jian F, Yongming T. Bistable behavior of the cylindrical origami structure with Kresling pattern. Journal of Mechanical Design. 2015;**137**(6):061406

[73] Pagano A, Yan T, Chien B, Wissa A, Tawfick S. A crawling robot driven by multi-stable origami. Smart Materials and Structures. 2017;**26**(9):094007

[74] Zhang Z, Chen G, Wu H, Kong L, Wang H. A pneumatic/cable-driven hybrid linear actuator with combined structure of origami chambers and deployable mechanism. IEEE Robotics and Automation Letters. 2020;**5**(2):3564-3571

[75] Bhovad P, Kaufmann J, Li S. Peristaltic locomotion without digital controllers: Exploiting multi-stability in origami to coordinate robotic motion. Extreme Mechanics Letters. 2019;**32**:100552

[76] Nishikawa J. Kresling Origami-Based, Magnetically Actuated Crawling Robot for Drug Delivery. The Ohio State University; 2021

[77] Mendoza MJ, Gollob SD, Lavado D, Koo BHB, Cruz S, Roche ET, et al. A vacuum-powered artificial muscle designed for infant rehabilitation. Micromachines. 2021;**12**(8):971

[78] Kulasekera AL, Arumathanthri RB, Chathuranga DS, Gopura R, Lalitharatne TD. A thin-walled vacuum actuator (ThinVAc) and the development of multi-filament actuators for soft robotic applications. Sensors and Actuators A: Physical. 2021;**332**:113088

[79] Oguntosin V, Akindele A. Design and characterization of artificial muscles from wedge-like pneumatic soft modules. Sensors and Actuators A: Physical. 2019;**297**:111523

[80] Kulasekera AL, Arumathanthri RB, Chathuranga DS, Gopura R, Lalitharatne TD. A Low-Profile Vacuum Actuator (LPVAc) with integrated inductive displacement sensing for a novel sit-to-stand assist exosuit. IEEE Access. 2021;**9**:117067-117079

[81] Weerasooriya LS, Chathuranga B, Somaratna OI, Kulasekera AL, Arumathanthri RB, Chathuranga DS, A novel contractile vacuum actuator and multi-actuator development for knee extension assist. 2021 IEEE 4th International Conference on Soft Robotics (RoboSoft); 2021

[82] Kulasekera AL, Arumathanthri RB, Chathuranga DS, Lalitharatne TD, Gopura RC, A low-profile vacuum actuator: Towards a sit-to-stand assist exosuit. 2020 3rd IEEE International

Conference on Soft Robotics (RoboSoft); 2020

[83] Bruzek R, Biess L, Al-Nazer L, Development of rail temperature predictions to minimize risk of track buckle derailments. ASME/IEEE Joint Rail Conference; 2013

[84] Singamaneni S, Tsukruk VV. Buckling instabilities in periodic composite polymeric materials. Soft Matter. 2010;**6**(22):5681-5692

[85] Chen D, Yoon J, Chandra D, Crosby AJ, Hayward RC. Stimuli-responsive buckling mechanics of polymer films. Journal of Polymer Science Part B: Polymer Physics. 2014; **52**(22):1441-1461

[86] Hu N, Burgueño R. Buckling-induced smart applications: Recent advances and trends. Smart Materials and Structures. 2015;**24**(6):063001

[87] Rogers JA, Someya T, Huang Y. Materials and mechanics for stretchable electronics. Science. 2010;**327**(5973): 1603-1607

[88] Wang Y, Yang R, Shi Z, Zhang L, Shi D, Wang E, et al. Super-elastic graphene ripples for flexible strain sensors. ACS Nano. 2011;**5**(5):3645-3650

[89] Shim J, Perdigou C, Chen ER, Bertoldi K, Reis PM. Buckling-induced encapsulation of structured elastic shells under pressure. Proceedings of the National Academy of Sciences. 2012; **109**(16):5978-5983

[90] Florijn B, Coulais C, van Hecke M. Programmable mechanical metamaterials. Physical Review Letters. 2014;**113**(17):175503

[91] Wang P, Casadei F, Shan S, Weaver JC, Bertoldi K. Harnessing buckling to design tunable locally resonant acoustic metamaterials. Physical Review Letters. 2014;**113**(1): 014301

[92] Yang D, Verma MS, So JH, Mosadegh B, Keplinger C, Lee B, et al. Buckling pneumatic linear actuators inspired by muscle. Advanced Materials Technologies. 2016;**1**(3):1600055

[93] Overvelde JT, Kloek T, D'haen JJ, Bertoldi K. Amplifying the response of soft actuators by harnessing snap-through instabilities. Proceedings of the National Academy of Sciences. 2015; **112**(35):10863-10868

[94] Robertson MA, Paik J. New soft robots really suck: Vacuum-powered systems empower diverse capabilities. Science. Robotics. 2017;**2**(9):eaan6357

[95] Hwang Y, Paydar OH, Candler RN. Pneumatic microfinger with balloon fins for linear motion using 3D printed molds. Sensors and Actuators A: Physical. 2015;**234**:65-71

[96] Mosadegh B, Polygerinos P, Keplinger C, Wennstedt S, Shepherd RF, Gupta U, et al. Pneumatic networks for soft robotics that actuate rapidly. Advanced Functional Materials. 2014; **24**(15):2163-2170

[97] Onal CD, Rus D, A modular approach to soft robots. 2012 4th IEEE RAS & EMBS International Conference on Biomedical Robotics and Biomechatronics (BioRob); 2012

[98] TolleyMichael T, ShepherdRobert F, GallowayKevin C, WoodRobert J, WhitesidesGeorge M. A resilient, untethered soft robot. Soft Robotics. 2014;**1**(3)

[99] Balak R, Mazumdar YC, Multi-modal pneumatic actuator for twisting,

extension, and bending. 2020 IEEE/RSJ International Conference on Intelligent Robots and Systems (IROS); 2020

[100] Ge L, Dong L, Wang D, Ge Q, Gu G. A digital light processing 3D printer for fast and high-precision fabrication of soft pneumatic actuators. Sensors and Actuators A: Physical. 2018; **273**:285-292

[101] Wang X, Zhou H, Kang H, Au W, Chen C. Bio-inspired soft bistable actuator with dual actuations. Smart Materials and Structures. 2021;**30**(12): 125001

[102] Yap HK, Ng HY, Yeow C-H. High-force soft printable pneumatics for soft robotic applications. Soft Robotics. 2016; **3**(3):144-158

[103] Zhang YF, Zhang N, Hingorani H, Ding N, Wang D, Yuan C, et al. Fast-response, stiffness-tunable soft actuator by hybrid multimaterial 3D printing. Advanced Functional Materials. 2019; **29**(15):1806698

[104] Peele BN, Wallin TJ, Zhao H, Shepherd RF. 3D printing antagonistic systems of artificial muscle using projection stereolithography. Bioinspiration & Biomimetics. 2015; **10**(5):055003

[105] Mutlu R, Tawk C, Alici G, Sariyildiz E. A 3D printed monolithic soft gripper with adjustable stiffness. IECON 2017-43rd Annual Conference of the IEEE Industrial Electronics Society; 2017

[106] Tawk C, in het Panhuis M, Spinks GM, Alici G. Bioinspired 3D printable soft vacuum actuators for locomotion robots, grippers and artificial muscles. Soft Robotics. 2018;**5**(6):685-694

[107] Stano G, Percoco G. Additive manufacturing aimed to soft robots fabrication: A review. Extreme Mechanics Letters. 2021;**42**:101079

[108] Rahim TNAT, Abdullah AM, Md A. Recent developments in fused deposition modeling-based 3D printing of polymers and their composites. Polymer Reviews. 2019;**59**(4):589-624

[109] Kamran M, Saxena A. A comprehensive study on 3D printing technology. MIT International Journal of Mechanical Engineering. 2016;**6**(2):63-69

[110] Lewis JA. Direct ink writing of 3D functional materials. Advanced Functional Materials. 2006;**16**(17): 2193-2204

[111] Khosravani MR, Reinicke T. On the environmental impacts of 3D printing technology. Applied Materials Today. 2020;**20**:100689

[112] Tee YL, Peng C, Pille P, Leary M, Tran P. PolyJet 3D printing of composite materials: Experimental and modelling approach. Jom. 2020;**72**(3):1105-1117

[113] Byrne O, Coulter F, Glynn M, Jones JF, Ní Annaidh A, O'Cearbhaill ED, et al. Additive manufacture of composite soft pneumatic actuators. Soft Robotics. 2018;**5**(6):726-736

[114] Stano G, Arleo L, Percoco G. Additive manufacturing for soft robotics: Design and fabrication of airtight, monolithic bending PneuNets with embedded air connectors. Micromachines. 2020;**11**(5):485

[115] Qi S, Guo H, Fu J, Xie Y, Zhu M, Yu M. 3D printed shape-programmable magneto-active soft matter for biomimetic applications. Composites Science and Technology. 2020;**188**:107973

[116] Al-Rubaiai M, Pinto T, Qian C, Tan X. Soft actuators with stiffness and

shape modulation using 3D-printed conductive polylactic acid material. Soft Robotics. 2019;**6**(3):318-332

[117] Robertson MA, Paik J, Low-inertia vacuum-powered soft pneumatic actuator coil characterization and design methodology. 2018 IEEE International Conference on Soft Robotics (RoboSoft); 2018

[118] Rus D, Tolley MT. Design, fabrication and control of soft robots. Nature. 2015;**521**(7553):467-475

[119] Sachyani Keneth E, Kamyshny A, Totaro M, Beccai L, Magdassi S. 3D printing materials for soft robotics. Advanced Materials. 2021;**33**(19): 2003387

[120] Zhou W, Li Y. Modeling and analysis of soft pneumatic actuator with symmetrical chambers used for bionic robotic fish. Soft Robotics. 2020;**7**(2): 168-178

[121] Moseley P, Florez JM, Sonar HA, Agarwal G, Curtin W, Paik J. Modeling, design, and development of soft pneumatic actuators with finite element method. Advanced Engineering Materials. 2016;**18**(6):978-988

[122] Joe S, Totaro M, Beccai L. Analysis of soft Kirigami unit cells for tunable stiffness architectures. In: 2021 IEEE 4th International Conference on Soft Robotics (RoboSoft). 2021

[123] Park S, Mondal K, Treadway RM III, Kumar V, Ma S, Holbery JD, et al. Silicones for stretchable and durable soft devices: Beyond Sylgard-184. ACS Applied Materials & Interfaces. 2018; **10**(13):11261-11268

[124] Moučka R, Sedlačík M, Osička J, Pata V. Mechanical properties of bulk Sylgard 184 and its extension with silicone oil. Scientific Reports. 2021; **11**(1):1-9

[125] Calderón AA, Ugalde JC, Zagal JC, Pérez-Arancibia NO. Design, fabrication and control of a multi-material-multi-actuator soft robot inspired by burrowing worms. In: 2016 IEEE international conference on robotics and biomimetics (ROBIO). 2016

[126] Ding L, Dai N, Mu X, Xie S, Fan X, Li D, et al. Design of soft multi-material pneumatic actuators based on principal strain field. Materials & Design. 2019; **182**:108000

[127] Fu H-C, Ho JD, Lee K-H, Hu YC, Au SK, Cho K-J, et al. Interfacing soft and hard: A spring reinforced actuator. Soft Robotics. 2020;**7**(1):44-58

[128] L-y Z, Gao Q, J-z F, Q-y C, Zhu J-p, Sun Y, et al. Multimaterial 3D printing of highly stretchable silicone elastomers. ACS Applied Materials & Interfaces. 2019;**11**(26):23573-23583

[129] Totaro M, Bernardeschi I, Wang H, Beccai L. Analysis and optimization of fully foam-based capacitive sensors. In: 2020 3rd IEEE International Conference on Soft Robotics (RoboSoft). 2020

[130] Joe S, Wang H, Totaro M, Beccai L. Sensing deformation in vacuum driven foam-based actuator via inductive method. Front Robot AI. 2021;**8**:742885

[131] Wang H, Bernardeschi I, Beccai L. Developing reliable foam sensors with novel electrodes. In: 2019 IEEE Sensors. 2019

[132] Nakayama R, Suzuki R, Nakamaru S, Niiyama R, Kawahara Y, Kakehi Y. MorphIO: Entirely soft sensing and actuation modules for programming shape changes through tangible interaction. In: Proceedings of the 2019

on Designing Interactive Systems Conference. 2019

[133] Wang H, Totaro M, Beccai L. Toward perceptive soft robots: Progress and challenges. Advanced Science. 2018; **5**(9):1800541

[134] Jiang Y, Wang Q. Highly-stretchable 3D-architected mechanical metamaterials. Scientific Reports. 2016; **6**(1):1-11

[135] Janbaz S, Bobbert F, Mirzaali M, Zadpoor A. Ultra-programmable buckling-driven soft cellular mechanisms. Materials Horizons. 2019; **6**(6):1138-1147

[136] Kaur M, Kim WS. Toward a smart compliant robotic gripper equipped with 3D-designed cellular fingers. Advanced Intelligent Systems. 2019;**1**(3):1900019

[137] Ge Q, Chen Z, Cheng J, Zhang B, Zhang Y-F, Li H, et al. 3D printing of highly stretchable hydrogel with diverse UV curable polymers. Science advances. 2021;7(2):eaba4261

[138] PROBOSCIdean Sensitive Soft Robot for Versatile Gripping. 2019. H2020-FETOPEN, Grant n. 863212. 2022. Available from: https://proboscis.eu/ [Accessed: January 24, 2022].

Section 4

Research of the Contradictory Problems

Chapter 6

A Systematic Study on TRIZ to Prepare the Innovation of 3DPVS

Haobo Yuan

Abstract

Regarding the innovation of biomimetic cell culture scaffold, 3DPVS, namely 3D printed vibratory scaffold, has been proposed as a present-to-future novel product. It currently stands at the stage of conceptual development. Design studies on 3DPVS Concept Generation show high value, and one essential part inside this could dwell at establishing design methodological knowledge that has innovation merits. TRIZ with its tools has proven value on creation and design innovativeness while they have not yet been utilized for scaffold design at mature level. In this paper, we attempt to study and explore the design aspects of TRIZ and its most relevant tools on the context of 3DPVS, as well as preliminarily indicating a TRIZ-based methodology, which could tailor the design aspects of 3DPVS. It also, to some extent, fills a gap in scaffold engineering and TRIZ literature and provides a comprehensive overview of a timely topic.

Keywords: 3D scaffold, conceptual design, engineering design, 3DPVS, product innovation, TRIZ, bio-design, artificial biomimetics

1. Introduction

An important product in bioengineering is 3D scaffold, which mimics the real tissue in vitro to achieve the external cultivation of cells. Design studies focusing on 3D scaffold have drawn increasingly attention in past decade, and increasing number of researchers show interest toward developing novel scaffold product. Inside scaffold innovation, 3DPVS, namely the 3D Printed Vibratory Scaffold, has been proposed as a present-to-future novel scaffold design, which currently stands at the stage of conceptual development [1]. To achieve the novel conceptual design of 3DPVS, TRIZ methodology with its essential tools will be explored in this study, hopefully contributing to a scientific-sound ground for future conceptual generation works using TRIZ. In this connection, the paper is structured as follows. Firstly, several background information will be introduced, including 3DPVS, its conceptual design process, traditional methodology for scaffold design with the limitations. Second, from main context, we will study TRIZ and relevant tools for 3DPVS innovation, the philosophy behind TRIZ, TRIZ contradiction-solving process. Several core tools inside TRIZ will be studied and analyzed in both general context and the specific one in 3DPVS design. This includes Altshuller Contradiction Matrix (ACM), TRIZ Innovation Principles, Substance-field modeling and Analysis (SFMA), Innovation standard solutions (ISS), ARIZ, and so forth. In the end, future work regarding ways of establishing TRIZ methodology will be discussed, as well as the proposals for validating the eligibility of applying TRIZ-based method from traditional engineering field into bio-design. The structure of this paper as well as the interconnection of each parts can be illustrated as follows:

Scaffold engineering and 3DPVS. Cell culture scaffold is defined as a class of artificially created biomimetic products used for culturing cells in vitro, through mimicking some real tissue properties. Scaffold engineering has developed in two directions. One is "from static into dynamic," and another from "2D into 3D" [2]. Existing scaffolds basically have a nature of being passive (also called "static"), which have several inevitable limitations. Taking this as well as the evolutional ladder of scaffold engineering into account, the novel concept 3DPVS (3D Printed Vibratory Scaffold) has been put forward, indicating that traditional 2D or 3D scaffold as the lower part of ladder while 3DPVS could stand at a relatively higher positionality. This was also justified via studying previous traditional scaffold as well as by penetrating that into the "Laws of System Evolution (LSE)."

In terms of the 3DPVS, concept indicates that a trinity of separate elements, namely vibration, scaffold, and 3DP, would turn into a unified systematic functioning with promising "vibratory properties" endowed by the scaffold itself, instead of through external mechanical ways such as connecting scaffold with a vibrator to acquire some vibrations on cell culture. Further, 3DPVS might concern with transforming the role of scaffold passively receiving vibration into potentially generating vibration at active or proactive way. In brief, localized vibratory function, as one core inside 3DPVS, was indicated as one of the most useful factors from evolutionary point of view; and 3D printing (3DP), another core, was pointed as the technology bridging traditional scaffold into future novel vibratory scaffold. Consequently, 3DPVS would possess the existing merits of traditional scaffold while innovatively providing tailored vibratory functions on different kind of external cell cultures.

Conceptual design process of 3DPVS. In author's previous work, where details about the conceptual process and design schematics were illustrated, conceptual design of 3DPVS could be proposed into three main stages, namely Design Initiation, Concept Generation, and Concept Evaluation [3]. The main task of Concept Generation phase is to innovatively create possible solutions or principles that help realize all or most desired functions designated for 3DPVS. Methodology and tools for generating innovation concepts compose the core of this stage [4]. In terms of Concept Evaluation, work of which is to create criteria, rank concepts, evaluate to what extent the generated concepts satisfy proposed requirements while providing feedback to design as a circular-improving process. To guarantee the quality of these two stages, one stage needs to be completed at relatively mature level, and it is called as Design Initiation, where the literature study, problem selecting and requirement identifying have been its chief focus. Regarding the methodology of each stage, Design Initiation chiefly takes Literature Review as its methodology for gathering all possible inputs and establishing relevant base models. For Concept Evaluation, tools such as Quality Function Deployment (QFD) to generate concept criteria, rank scores, and select optimal concept could be used. Experimental or computer simulation method would be useful regarding evaluation. It is perceivable that only after a relatively thorough work of studying and establishing methodology, it could logically come to an effective concept generating work followed by near future design.

Traditional methodology for scaffold conceptual design. Through previous literature study on two areas, namely the area of design methodology and the area of 3D scaffold's design, it reveals that there might exist a gap between these two fields [5].

Using design methodology, such as TRIZ, for instance, it has been already applied for a wide range of research fields, including electronic product design, mechanical objects innovation and industrial designs, and so forth, but little studies have investigated it for the design of 3D scaffold. Reasons could partially lie on the fact that firstly traditional scaffold engineering is not very demanding when novelty or innovation is concerned, so changes on scaffolds can be acquired through CAD software or experimental tests; and secondly, TRIZ researchers probably did not draw much attention on scaffold innovation. In this connection, a formal design methodology has not yet been applied for the design of novel scaffold. Although several attempts have been made by using some part of QFD, TRIZ, or Axiomatic Design, the work generally remains immature, and much more effort in this direction is needed [6–8]. The typical works during the scaffold development, such as using CAD software or tools to modify parameters based on existing old models, as well as through experimental "test and error" method to investigate proper properties of scaffold then make relevant changes, might largely be considered as design approach while neither can be viewed as formal methodology.

At last, it is important to mention that selected methods or tools should be convenient for engineers without strict background on designs. Due to the complicated and cross-domain properties of 3DPVS, as an already-proven fact, it might be not practical, or effective, to use another complicated system for designing it. Therefore, the approach of this paper is expected to provide an all-in-one, fully, or to large extent, TRIZ toolset where future research studies on conceptual design of 3DPVS could directly be benefited with convenience and feasibility.

2. Selecting TRIZ for 3DPVS innovation

2.1 Innovation as the essence of 3DPVS design

Following the laws and principles of scaffold evolution, it is reasonable to reach the conclusion that traditional 3D scaffold system, vibration system, as well as cell culture system, will develop simultaneously and meet at a crossing-point at near future, which gives the indication for the appearance of novel 3DPVS. There have been no previous studies that concerned 3DPVS, either on its conceptual philosophy or product realization. In this connection, a proper methodology, or a methodological system needs to be selected, adapted, or even invented in first place to help achieve the 3DPVS at conceptual design level.

Comparing with traditional scaffold, the development of novel product 3DPVS is an innovation process, which is different from that of other bio-products, which can be better approached by experimental "trial and error" strategy from beginning. Experimental focus at this stage of 3DPVS is considered as less vital compared with a conceptual design at philosophical level. In other words, the former can be useful and efficient only after a success approach of the latter. Also, the core task of 3DPVS conceptual design revolves around the idea "innovation," which means that creating new concepts where the newly proposed vibratory functionality could innovatively integrate with the 3DP fabricability and 3D scaffold. This integrating process is novel, and it is more promising to be achieved when selected method that especially tackles with this "novelty." TRIZ therefore can be potential since it was especially developed with the blueprint for "creativity" and "innovation."

Compared with TRIZ, another innovation method is Extenics, which focuses on solving incompatible problems by formularized methods in management and engineering. Extenics has solved many engineering incompatible problems and has been practiced as suitable for special fields. However, there is no systematic theory on knowledge-based innovation [9]. Extension innovation methods need further improvement to be used as general operable methods for innovative activities. Regarding bio-innovation, TRIZ shows more applicability. Therefore, TRIZ is considered as better approach for 3DPVS scenario.

On other hand, as studied early [2], tackling contradiction in GMBV (geometric-, mechanical-, biological-, and vibratory-) characterization of 3DPVS, increasing ideality level of 3DPVS to gradually approach its ideal model as well as viewing the whole design process from the idea of system evolution, could be several essential focuses that the selected methodology needs to fulfill. In this way, TRIZ methodology has been identified as highly proper since inside TRIZ the "contradiction matrix," "innovation principles," and "evolution patterns for ideality" exactly tailor the requirements of those three cores, which can be also the consideration criteria when selecting design method. Considering these, TRIZ will be used for the innovation process of 3DPVS development. Contents of TRIZ and its innovation mechanisms will be studied in following section.

2.2 TRIZ innovation mechanisms

TRIZ, namely "Theory of Inventive Problem Solving," a is set of tools used by engineers, inventors, and scientists to systematically create new innovative ideas and solutions. It has been used on analyzing and studying artificial engineering systems [10, 11]. Instead of waiting for random inspiration or using trial and error to innovate, TRIZ enables designers to attain breakthrough ideas, methods, and solutions on demand. Created in Russian, Tools of TRIZ have been developed from the study of millions of patents. Rather than playing psychological games, brainstorming, or guessing-and-trying as in ordinary designs [12], TRIZ solutions are systematically derived based on the information abstracted from studying how others solved problems and created innovation.

TRIZ methods, as discussed, have been created, tested, and analyzed based on millions of patents and innovations, aiming to transform the subjective innovation process, to some extent, into a relatively objective, scientific, and easy-handling one. Basic philosophy of TRIZ indicates that, when innovation problem is broken into required elements inside TRIZ as input, following a series of "alchemical work," the being of the original problem filled with contradictions, splits, and conflicts, can turn into a new being, which is of integration. In brief, TRIZ analyzes the hidden determinants of creative works done by previous innovators and packages the essence of which into TRIZ. Technical and physical contradictions are the cornerstones of TRIZ. Traditional TRIZ generally contains three mechanisms as follows (**Table 1**) [13–15]:

To be specific, the categorization of mechanisms depends on the essence type of the contradiction. Process of using the contradiction matrix to solve a problem occupies the most significant role in TRIZ methods [7, 13, 16]. In this connection, system contradiction (technical contradiction) is the conflict between two or more parameters inside a system, while physical contradiction is the conflict between two or more values of one parameter. A technical contradiction could mean when improving one design parameter of the product, the other design parameter will deteriorate, so improving one parameter could be conflicting with the properties of others. In the design of 3DPVS, for instance, when the pore size is increasing to a desired level to cooperate with vibration, the geometric stability may decrease and negatively affect cell culture process. Thus, the elimination of this conflict could indicate that designers need to find

Problem-solving mechanism related with system contradiction
Thirty-nine parameters and forty Inventive Principles can possibly be applied. When it is Software Contradiction, then use 24 parameters and 40 Inventive Principles.
Problem-solving mechanism related with physical contradiction
Five Separation Principles can be used for solving Physical Contradiction.
Problem-solving mechanism related with non-contradiction issues
The System of Standard Solutions will be utilized to address problems where no contradiction can be abstracted or where the contradiction is too challenging to be categorized by either System or Physical Contradiction

Table 1.
Mechanisms of TRIZ for problem-solving related with different contradiction cases.

a solution that fulfills both requirement of scaffold's pore size for possible vibration and avoiding the deterioration property of its geometric stability. In contrast, a physical contradiction of which could mean that when pore size changes to a desired level, say, highly precise at nano-micro level, the fabricability of this pore at such size deceases or even becomes impossible, so how to identify the proper size of pore, being both precise as required and with its fabricability ensured, will be the key in this scenario.

As far as system evolution is concerned, it is guided by the defining and resolving System and Physical Contradictions in terms of performance demands. Software Contradictions are modified from System Contradictions tailored for software and informational-technology-related problems [17]. If there is contradiction, then based on its nature, that is, whether it is physical contradiction or technical one, the problem can be approached using either the Five Separation Principles (physical) or Contradiction Matrix and 40 principles (technical). If no contradictions can be recognized or contradiction is at beyond level, tools such as evolution principles and laws, 76 standard solutions, etc., can be potentially applied to solve the problem. To conclude, the process of using Contradiction Matrix can be briefly summarized as in **Figure 1**.

For another instance, when enhancing the print resolution of 3DP, we usually need more printing materials, and the resource consumption of 3D printer will thus increase. The conflict between printing resolution (to increase) and material consumption (to decrease or preserve) requires that both "more and less" printing materials should be used, which is a technical contradiction. In terms of customer aspect, 3D printer tends to be complex and sophisticated equipment and more operations are required to maintain the functions of printer. However, customers might be unwilling to spend excessive time operating this, and they prefer to use simple, easy ones while these could not fabricate the fine-resolution samples. Low maintenance and fine resolution therefore become a system contradiction. In terms of physical contradiction, enhanced resolution does not always lead to a high-quality printed sample. Cells might not pass through fabricated scaffold if the resolution is too fine. A resolution, neither too fine nor too coarse, leads to a physical contradiction.

2.3 Possible cons and pros of applying TRIZ on 3DPVS design

Despite the potential benefits of TRIZ for product innovation and novel design, there inevitably exist some limitations of TRIZ. Identifying these limitations at first place could help provide designers with better clarity of TRIZ in terms of pros and cons. This section we will study two aspects, that is, firstly, the limitation of TRIZ in general engineering, and secondly, the possible challenges when using TRIZ for specific scaffold design.

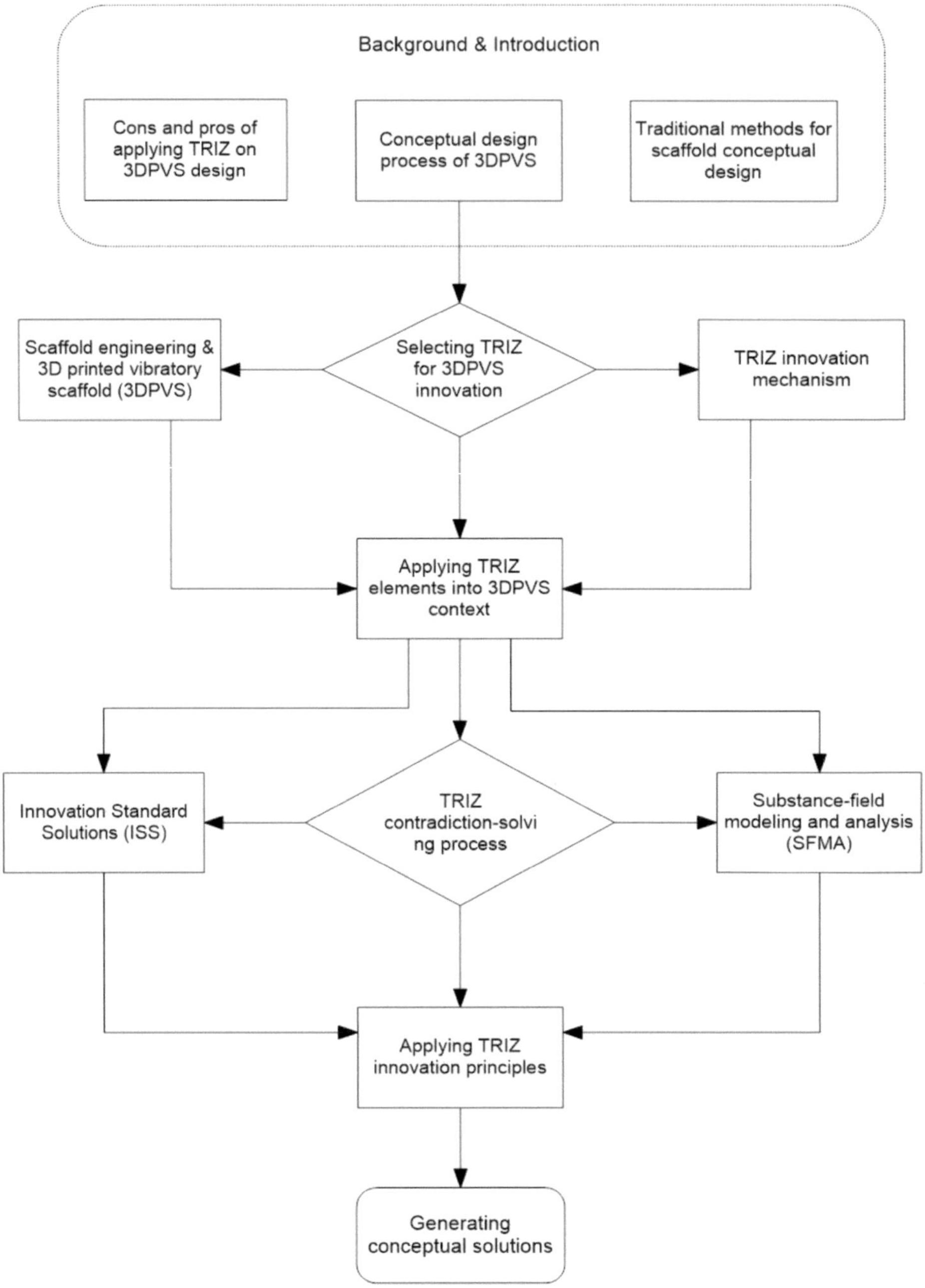

Figure 1.
Paper structure and potential interconnection of each section.

2.3.1 Possible limitations of TRIZ in general

Based on the literature reviews on TRIZ and TRIZ-design case studies, the limitation is chiefly recognized by its difficulty of utilization in several aspects. This includes firstly translating problems or innovation issues from specific into generic, secondly, identifying TRIZ-based solutions for these generic problems, and thirdly, translating the generated solutions back to the specific problems or issues. In terms of using TRIZ innovation principles or standard solutions, a careful work and thorough preparation of "translating" them into the context where product innovation can directly take advantage are always necessary [18–20]. This usually takes much time and effort. Though TRIZ innovation principles and philosophy remain the same for every designer, the expertise or understanding of which varies from one to another. Lack of efficiency and precision in this translating work therefore can be an obstacle for those who attempt to TRIZ on innovation designs. Also, it tends to be challenging when applying TRIZ for innovation in bioengineering scenarios, since TRIZ has not been explored and utilized for this context mainly.

From another perspective, first, it is rare to find previous studies regarding the utilization of TRIZ methods on novel scaffold innovation. Although TRIZ methods have been used and approved by researchers in the miscellaneous, knowledge of TRIZ on scaffold innovation is lacking. To be specific, there lacks formal research in terms of analyzing contradictions and selecting principles for scaffold innovation. Considering this, a TRIZ-based system for scaffold design is therefore needed, aiming to translate TRIZ principles into scaffold-tailored scenarios specifically for 3DPVS. On other hand, TRIZ methodology has been used cooperating with other proven supportive design methods, such as Quality Function Deployment, Root Cause Analysis, System Function Analysis, Hybrid System Design, FMED, and Trimming, etc. [5, 21] . There lacks a systematic study regarding how to link these proven supportive methods on scaffold innovation cases. Therefore, a thorough process combining TRIZ and these supportive methods for scaffold innovation remains as a gap and is therefore of urgency to be developed.

2.3.2 Advantages of TRIZ for scaffold innovation

Comparing TRIZ methodology with traditional methods inside Engineering design process (EDP), which is a series of steps that engineers follow to generate solutions to a definite design problem, there exist several advantages. Traditional ways of generating product design concepts or solutions could base on the process such as Brainstorming, "Trial and Error" methods, and so on [22, 23]. Though such approaches are easy to implement, they tend to lack the innovation essence resulted from core of creativity [24]. In addition, following traditional problem-solving path, which can be considered as "linear" [25], we might find something negative and start making efforts to alter it; we may succeed in obtaining a certain result, but together with this result another result might occur, which we did not in the least expect or desire and which we could not have suspected. Using cell scaffold, for example, an additive added to scaffold could directly help cell adhesion, but this may cause unknown proliferation effects, which designers could not predict previously. Correcting this usually means re-conducting the cell culture experiment, which makes the whole process repetitive, costly, and time-consuming. Due to this fact, a more efficient approach in terms of generating innovative, useful, and effective concepts for scaffold design has attracted much attention. In this connection, TRIZ has been proposed as right method, and the four-route process is established.

On the other hand, for a scaffold engineering system, everything within could be interconnected and every function might be inevitably counterbalanced by some other function or by a whole series of other functions, though they could not be easily detectable in linear patterns. In this connection, the new path to be developed should not be linear, but dimensional, which means that the possible supplementary changes, or side effects, as far as possible contradictions inside GMBV is concerned, need to be considered beforehand. Since TRIZ allows a nonlinear contradiction-solving path, it could logically help designers become better aware of the interconnection of the various properties within the cell culturing scaffold system. Further, due to the rather complex, sophisticated design issues for 3DPVS as bio-design and cross-domain, ARIZ inside TRIZ could be highly suitable tool since it was specially developed for complex design questions. Introducing ARIZ into complex bio-product innovation thus could possibly break a new ground.

2.4 Transferring TRIZ elements into 3DPVS context

The innovation work at current stage of 3DPVS needs to focus on its conceptual design. To ensure a high-quality innovation process, a systematic methodology is of significance [26]. In this connection, TRIZ has been identified as potential methodology for 3DPVS innovation, and in the following sections, we aim to tailor the traditional TRIZ method and the related tools into a systematic methodology which can be specified for the conceptual innovation and design of 3DPVS. Using this methodology system, designers could possibly generate required solutions and concepts for 3DPVS. To establish such methodology, work is structured as follows. First, TRIZ and its basic mechanisms will be introduced, with the justifying work why TRIZ would be an optimal means to help accomplish the conceptual innovation of 3DPVS. Following this, it comes to the core part of the research work, which is to tailor and adapt original TRIZ-based methods into structured methodology, which tackles with the innovative tasks inside the conceptual design of 3DPVS. Since the nature of 3DPVS currently remains as a future novel concept and that its reality product has not yet been approached by utilizing TRIZ, major proportion of elements inside TRIZ consequently need to be transferred, at least to some extent, into a set of specific languages, codes, or forms that can be straightly applicable on novel scaffold design. Besides the novelty of 3DPVS, the very process of this transferring might also be considered as a novel aspect.

3. TRIZ contradiction-solving process for 3DPVS context

For TRIZ, systems evolve toward gaining ideality by overcoming contradictions. Conceptual solutions of 3DPVS would be the output of TRIZ Innovation process, and vital part of this innovation process is achieved by identifying "contradictions" then solving them [27]. Contradiction-solving is the unique mechanism of TRIZ, which might be considered as an evolutionary aspect advantaging other design method.

In this research, large part of work in conceptual development of 3DPVS would revolve around the "center of gravity," namely "contradiction-solving," which is achieved through generating identified contradiction and conflict pairs and solving them in a scientific manner. The philosophy behind could be that the innovation and problem-solving process is the process of eliminating contradictions, which is achieved through right intensification of conflicts, not by smoothing them or using buffers to "avoid seeing." In this connection, the contradiction solving process will potentially penetrate every stage regarding the conceptual scenario of 3DPVS.

3.1 TRIZ contradiction-solving into scaffold innovation

A systematic approach regarding how to apply TRIZ in 3DPVS design. Compared with traditional single-direction problem-solving by TRIZ, in this study we expand a four-route process, which is better tailored for sophisticated design scenarios. Since the very nature of 3DPVS being miscellaneous, the innovation process requires cross-domain knowledge and expertise, which makes the 3DPVS innovation relatively complex and sophisticated compared with single-domain product innovation. In brief, the innovation of 3DPVS firstly requires integrating three core set contradictions, that is, the contradictions between 3D scaffold and 3D printing, between 3D scaffold and vibration mechanism, and between vibration and 3D printing; secondly, the set of pair-based contradictions inside GMBV characterization of 3DPVS, namely the contradiction between relations of G-B, G-V, B-V, B-M, G-M, M-V, needs to be carefully addressed.

Figure 2 below shows the scope this Contradiction Resolving Approach of six-set contradictions between four design aspects, that is, geometric, mechanical, biological, and vibratory properties. It is worth noting that, besides these six contradiction relations, for each G-, M-, B-, V- characteristics inside 3DPVS design, they could produce conflicts within to contradict with themselves, and this will create four sets of most typical physical contradictions, that is, G-G, B-B, M-M, and V-V contradictions (**Figure 3**).

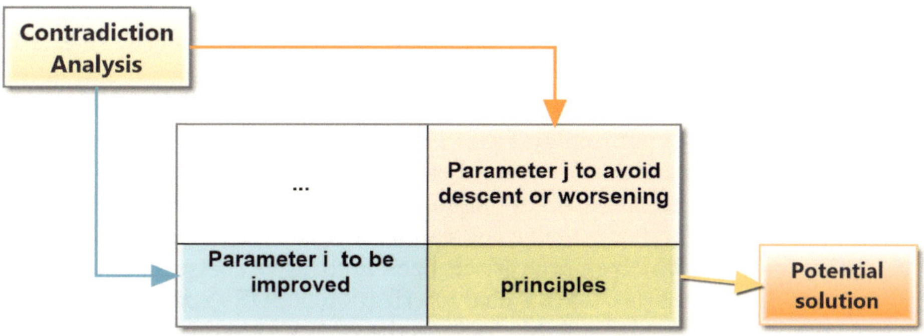

Figure 2.
The process of utilizing TRIZ contradiction matrix.

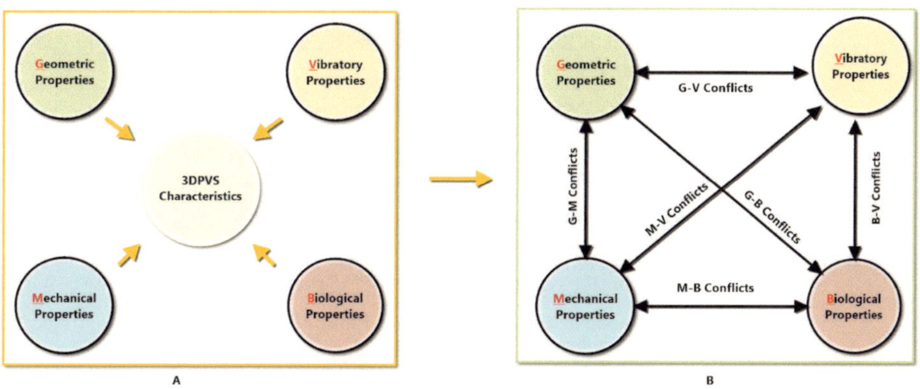

Figure 3.
Six relation sets of contradictions existing for 3DPVS design.

3.2 ACM applying on 3DPVS conceptual design

As essential part of TRIZ, ACM helps solve System Contradiction (SC), where improving one parameter of a system conflicts with the requirements of other parameters. To define an SC for system and find relevant parameters consisting of SC, four steps are required, and to select effective Inventive Principles for this contradiction, another four steps will be needed, which makes up an eight-step process for the whole contradiction defining and solving. In this connection, we could embed the 3DPVS design into this eight-step process as follows (**Table 2**).

For step 3, the question comes as which pair of the contradicting parameters ought to be analyzed as priority. The possible ranking order will depend on requirements and conditions of the given cell culture case scenario, and the different aims in different stage of 3DPVS design would also give different selecting priority.

3.3 Corresponding characteristics of 3DPVS with list of TRIZ parameters

To have a better understanding regarding 3DPVS parameters, it is important to know what parameters TRIZ summarizes as most vital ones. The traditional TRIZ uses 39 parameters for Engineering and Technology, and for 3DPVS context, predominant majority of them could be appropriately applicable.

At stage of Design Initiation, relevant parameters for 3DPVS would be collected and analyzed regarding the possible contradicting interconnection between each of the two parameters. For the selected 3DPVS parameters, which compose a System Contradiction, they need to transform from commonly used attributes of 3DPVS into the general parameters recognized by TRIZ. Further, here might be gaps between the commonly used parameters for scaffold engineering and the TRIZ parameters generated universally possible realm of engineering and science. Parameters of latter tend to be more generic and conclusive and could be directly utilized together with the 40 TRIZ principles, which exist as generalized innovative abstracts or concepts for triggering innovation. In brief, GMBV characterization of 3DPVS needs to transfer into TRIZ language at first place. **Table 3** shows the GMBV types into potentially corresponding TRIZ parameters.

Step No.	Tasks
1.	Generate the parameter list for the 3DPVS system, the detailed process of which could be generated in Design Initiation stage.
2.	Select a parameter at 3DPVS system level and change its value.
3.	Analyze the interconnection between the selected parameter and other remaining parameters; select out conflicting pairs, each of which will represent a System Contradiction.
4.	Choose the most appropriate parameters in the 39 parameters list, which correspond to the 3DPVS system parameters as selected above.
5.	Utilizing ACM, from the vertical list identify 3DPVS parameter whose value is to be improved.
6.	Identify the worsening or descending parameter inside the horizontal list.
7.	Identify the inventive principles from the matrix intersection.
8.	Study these principles in detail and analyze the applicability of each on 3DPVS design.

Table 2.
Eight-step process for the 3DPVS contradiction defining and solving.

3DPVS GMBV Characterization Type	Explanation	Potentially corresponding parameters of TRIZ
G-	1. Geometric Parameters (Pore characterization, porous connection)	1, 2, 3, 4, 5, 6, 7, 8, 12, 18, 25
M-	2. Material Composition (Chemical Parameters)	2,5, 7, 9, 36
M-	3. Mechanical parameters	3, 9, 14, 17, 19
B-	4. Biological parameters	5, 9, 17, 20, 28, 30,
V-	5. Dynamic/vibratory functionality parameters	2, 4, 6, 12, 26
V-	6. Vibration mechanism parameters	17, 19, 25
Others	7. 3DP Fabrication parameters	27, 32, 34, 35
Others	8. CAD Design and computer simulation	28, 38

Table 3.
Translating 3DPVS's GMBV attributes into the general parameters of TRIZ.

3.4 Optimizing innovation principle selection from Altshuller's contradiction matrix for 3DPVS

In terms of the Alshuller's Contradiction Matrix (ACM) for 3DPVS context, a multi-level principle selecting mode can be used, compared with traditional ACM, which uses single-level one-to-two principle selecting mode. That is, traditional ACM selects one improving parameter and the decreasing another, which causes confliction, then manually searches the 40 innovation principles and gets recommended solutions, which tend to be a combo of 3–5 principles. For such method, several concerns exist. Firstly, there is no ranking or evaluating tools to judge the weight of the selected 3–5 principles; designers need to compare one with another by subjective experience, which probably makes the process time-consuming and inaccurate. Secondly, for the pair of parameters creating conflicts, each of which might be a combined or resultant parameter from several other parameters. Simply choose one parameter for "X" axis in ACM and another for "Y" would be inadequate. In other words, If the parameter "X" contains three elements, say X1, X2, and X3 parameters as a unit, and Y parameter to decease includes another three elements, say Y1, Y2, and Y3, then traditional ACM could find it challenging in solving this. Starting from this new Matrix-searching process comes into the Innovation Principle Analysis (IPA), which would contain the similar process of selecting generic parameters in the matrix through fitting with the parameters composing System Contradiction. Designers will obtain the inventive principles and the associated percentage ranking, which is an special trait of the new ACM, in order to help solve contradiction effectively. In this connection, each pair of parameters, which create conflict, would be possibly divided into one, two or three sub-parameters, based on the parameter evaluation criteria; that is, parameter weight calculated by specific design requirements as into four levels, respectively representing by symbol "S," "A," "B," and "C." For the IPA in 3DPVS context, we would use "S," "A," where S is for the chief feature parameter in given scenario, A for the secondary, and B for minor. "C" is considered as trivial determinant, which does not affect the decision-making in innovation process. If the conflicting parameter is the exact one inside the 39 parameters list, IPA Matrix can be used similarly as traditional Matrix, with one difference, that is, weight of principle by percentage among other principles would be calculated as well. For this instance, as well as transforming traditional ACM into new ACM-IPA, example is shown in **Figure 4**, which also illustrates the basic development from traditional single-dimensional Matrix into the Novel multidimensional one.

Inside IPA, two or three sub-parameters constituting the states or functions of the original parameters will be used for filling the X or Y ACM column before the principle searching starts. In this connection, the total searching time would be a number between nine to one; a proportion ratio based on the appearance frequency of selected principles in this one-to-nine scenario will be calculated. The Company Time to Innovate has used such calculation philosophy in their TRIZ application and proved higher design efficiency. Top principles resulted will the considered as prioritized solutions for the given scenario. For better understanding this as 3–3 Matrix, an example is given here regarding a specific 3DPVS scenario. In this scenario, cells move in fast speed in a culturing environment as the chief feature parameter (S), the definite duration (A) needs to be ensured as well as maintaining the cell temperature (B). In one fast-moving state, cells create negative pressure on the scaffold, making the stress or pressure as the chief feature parameter (S) of Descending Parameters (DP). Also, more energy could be used by such higher movement by cells so other cell process might be affected, making the Energy Lost as the secondary feature (A); the shapes of cells might also be negative affected, which will result in undesired morphological effect. Shape issue is not vital for this scenario, but also it is a negatively affected factor, so to make it as the support feature parameter (B). For designers, it requires how to ensure the Ascending Parameters (AP), namely speed of cells, speedy period, and cell temperature, while not descending the DP pressure, energy use/lost, and shape? In this instance, the IPA result from ACM is shown in **Figure 5**.

From this IPA, we could see, principles 19, 35 represent the highest weight and principles 15,14 and 6 follow gradually afterward. These principles would be considered as the most promising direction when solving the regarded problem as brought up, while the principles weighting less could be used as reference to help innovative thinking. In this case, a possible innovative suggestion is that dynamic materials could be used so scaffold's physical and chemical parameters would change at periodic level. From this direction, we might be able to explore further in details.

To conclude, IPA is proposed to play vital role inside the four-route TRIZ-based methodology. In next section, we will illustrate the detailed knowledge about TRIZ

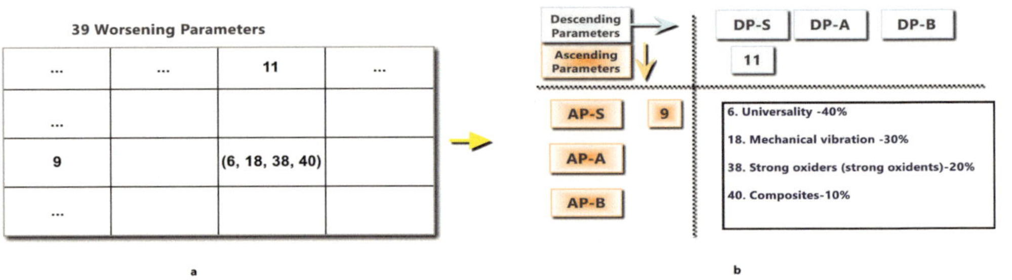

Figure 4.
From traditional 1–1 matrix solution (a) into the developed 3–3 matrix IPA (b), as well as the IPA results for traditional 1–1 conflict.

DOI: http://dx.doi.org/10.5772/intechopen.101576

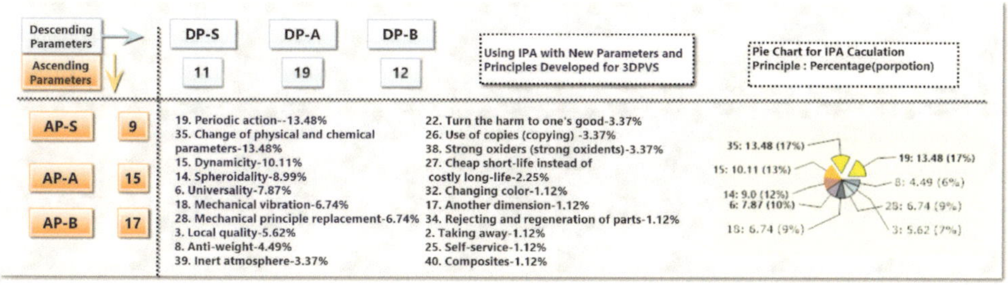

Figure 5.
IPA matrix illustration for solving a complicated 3–3 parametric cell culture issue.

innovation principles, as core contents for IPA and TRIZ, as well as discussing the most likely direction when applying each principle for future 3DPVS design.

4. Applying TRIZ innovation principles for 3DPVS design

After introducing IPA and its philosophy, in this section we would study the details of IPA and the most possible manifestation of each principle into 3DPVS design context. Firstly, the significance of translating TRIZ Innovation Principles will be discussed, secondly the 40 innovative principles, principle-related expansions, as well as 3DPVS design strategies utilizing corresponding principles will be illustrated.

4.1 Necessity of translating innovative principles into 3DPVS context

One core element of TRIZ is to recommend innovative principles, which potentially stimulate the designer's creative thinking in solving design conflicts or contradictions. However, TRIZ-based inventive principles have not developed for bioengineering products, neither have they utilized for scaffold innovation. This means that existing principles might not be applicable to scaffold design including conceptual development of 3DPVS. Therefore, the key of identifying a TRIZ-based innovative solution for 3DPVS lies on translating a set of scaffold-based conflicts into the generic conflicts established in TRIZ. Simultaneously, the potential solutions, inspired from innovative principles, need to be translated into solutions tailored for 3DPVS scenario.

In this connection, a wide degree of analysis, interpretation, and new understanding on TRIZ innovative principles are necessary when utilizing TRIZ on 3DPVS design. So to speak, the process of translating solutions from generic TRIZ-based traditional ones into the specific, tailored solutions applicable for scaffold design specifically for 3DPVS, will be of highly significance and contents of which will be discussed in the following section.

4.2 Innovative principle analysis (IPA) and 3DPVS potential strategy

TRIZ-based techniques traditionally use 40 inventive principles to eliminate System Contradictions (SC). How to utilize these innovative principles on 3DPVS tends to weigh significantly regards its proportion inside entire design process. However, for the novel cross-domain aspects existing in 3DPVS, old explanations could be limited, and new interpretation tailored for novel bio-design seems of

Separation principles	Indication	Corresponding innovation principles
Separation in Space	Object has parameter value A and B at Time "1" and "2" respectively.	1,2, 3, 4,,7, 13, 17, 24, 26, 30
Separation in Time	Object has parameter value A and B at Space "1" and "2" respectively.	9,10, 11, 15, 16, 18, 19, 20, 21, 29, 34, 37,
Separation upon conditions	Object has parameter value A and B on condition "1" and "2" respectively.	12, 28, 31, 32, 35, 36, 38, 39, 40
Separation between the Parts and the whole (system and sub- or super-system)	Object has parameter value A at system level, while value B at super-system level (or A at system while B at sub-system level)	1, 5, 6, 7, 8, 13, 14, 22, 23, 25, 27, 33, 35

Table 4.
Relationship between separation principles and 40 innovation principles.

urgency to explore. Aside from understanding 40 principles from traditional engineering perspectives [28], perception on 3DPVS based on bio-design especially for 3DPVS can be explored. In this connection, examples from general engineering world could be illustrated, in order to help clearly understand the definite principle.

4.3 Separation principles for solving physical contradictions of 3DPVS

While System Contradiction (SC) for 3DPVS is a conflict between two parameters at system level, namely different parameters either at 3DPVS's super-system, system, or sub-system levels, Physical Contradiction (PC) is the conflict between two desired values of one parameter at same 3DPVS system level. These PCs could be eliminated by four separation principles as follows in **Table 4**. In terms of the relationship between Separation Principles and the 40 TRIZ innovation principles, we could also categorize the latter into four classes. This categorization fully suits with 3DPVS product.

Except from direct analysis on PCs, transition from SC into PC could help understand SC as well as transforming it to an easier-level problem for solving. For sake of efficiency in 3DPVS design, SC to PC for example can be coded as:

{A higher scaffold contains more cells, WHEREAS has poor stability during dynamic cell culture.

Two conflicting parameters: 'height' of scaffold <– > 'stability' of the scaffold.

Physical Contradiction: One parameter, "height" of scaffold, ought to have different values of low and high.}

5. SFMA into 3DPVS scenario

Substance-field modeling and analysis (SFMA) is one vital part composing TRIZ engineering. SFMA is potentially of high value to the four-route methodology established previously. In this section, several aspects regarding SFMA as well as its application on 3DPVS will be studied as follows.

5.1 Studying SFMA and its philosophy

Invention and Separation Principles inside the TRIZ methodology, as discussed, could be applied for 3DPVS design. Cooperating with this, another tool inside

TRIZ for innovation and problem-solving is substance-field modeling and analysis (SFMA), which aims to provide standard solutions not related to specific areas of technology but enable the solution transferring from one scientific branch to another. This system, namely System of Standard Solutions, is especially beneficial for solving complicated problems by using a combination of several standard solutions. Philosophy of which is built upon the foundation that in different domains of science and industry there exist a definite number of graphic models that describe the oceanic amount of problems, thus a definite number of transformed graphic models can illustrate possible solutions. Each Standard Solution will propose one pair of such graphic models for solving similar, standard problems when System or Physical Contradiction in Route 1 and 2 could not be effectively anchored. In brief, Substance-filed Analysis (SFA) is the unique language of this system. For tailoring the novel 3DPVS design, SFR could be potentially utilized inside the design algorithm with several steps, which will trigger innovation and help generate fine ideas.

5.2 SFM and three-class analysis

5.2.1 Studying SFM

SFA contains three elements in chief, namely S1, S2, and F, the representation of which and possible application examples on 3DPVS design are shown in **Table 5**.

The analysis on two SFM, one of existing problem and another for its possible solution, is called Substance-Field Analysis. Study of the evolution of substance-field structures is also a part of SFA. This model illustrates the situations, problems, challenges, limitations, and possible solutions in a graphic, abstract form with three basic components, as shown in **Figure 6a**.

5.2.2 A three-class analysis for function between S1 and S2

Understanding the basic components of SFA, it is important to have a clear picture regarding the interconnected relationship between these three components, namely the relation in S1, S2, in S1 and F as well as in S2 and F. Three classes could be summarized corresponding to the relations, with three definite symbols representing different types of functions, shown in **Figure 6b**.

SFA elements	Description	3DPVS application example
Substance S1	a "tool" or "instrument" utilized for producing a product, manipulate a function, control, measure or change values of product's parameters;	materials and solid items like water, cell, cell culture medium, scaffold, vibrator, 3D printer etc.
Substance S2	the "product" or "object" which is to be produced, manipulated, measured or changed;	materials and solid items like water, cell, cell culture medium, scaffold, vibrator, 3D printer etc.; S1 in one design scenario could be S2 in another;
Field F	the energy or energy field or medium utilized for the interaction between S1 and S2;	energy like magnetic, mechanical, electrical, chemical, thermal, acoustic field etc.

Table 5.
SFA elements, description and 3DPVS application examples.

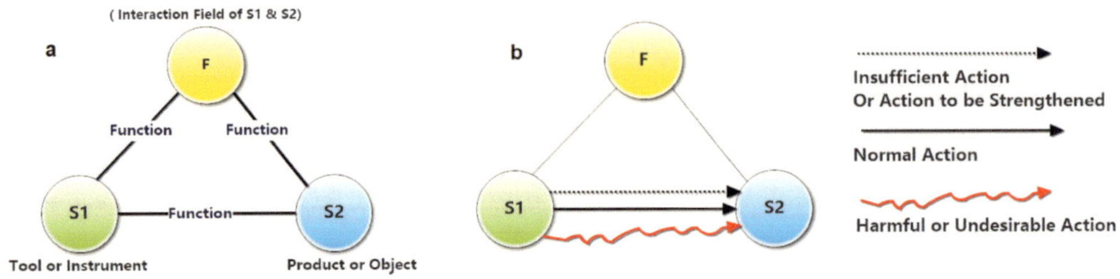

Figure 6.
Illustration of SFM containing three elements and three functions. a, SFM and representing symbols for function between substances. b.

For the three classes inside the SFA, that is, the class concerning system with incomplete or inadequate element or function, the class concerning system to eliminate harmful or hindering actions, and the class where system is to strengthen insufficient action, they have been summarized as **Table 6**, which also illustrates proper application on 3DPVS.

Class No.	System Indication	Description	Direction for 3DPVS scenario
Class 1	system with Incomplete or Inadequate SFM	Merely S1 exist and the system lacks F and S2, therefore S1 could not generate the Desired Functions. To solve this issue, F and S2 can be introduced to create a complete model;	Take S1 as Scaffold for instance. S1 is to have vibratory Functions, so Field or Vibratory Filed F provided by S2 needs to be introduced to provide the Vibratory Functions to S1. F can be a magnetic field and S2 the magnetically vibratory materials.
Class 2	System to Eliminate Harmful or Hindering Actions	The sub-field model is complete but exists harmful, negative actions. Designers need to integrate new substance or field to eliminate, counterbalance that negativity. That is, S1 here has normal and negative actions to S2 through original field F1. Solution might involves introducing new elements into system, i.e., S3 and F2 having positive or counterbalancing effect on S2 is introduced so original harms will cease to some extent.	For undesirable vibration frequency: S1 is vibrator, F1 the mechanical force to transmit the vibration and S2, the scaffold to be vibrated. Existing vibration is rough and imprecise; the function is not desirable since we want precise and subtle vibrations. Then a subtle vibration mechanism, a vibrator with higher controllability in frequency can be introduced as S3, and F2 will be a more subtle and controllable frequency effecting S2 scaffold.
Class 3	System to strengthen or enforce insufficient Action	Model is complete with S1, S2 and F, but effect from S1 onto S2 through F is deficient or inadequate. Solution for this comes as introducing new Substance S3 and Field F2, so the combined effect from S3 and F2 act as the Neutralizing Force which strengthens the original Deficient Action.	Scaffolds made of Smart Materials may lack strength on mechanical properties, i.e. the structures remain unstable and easy to break: Smart materials as S1, mechanical influence as F and Scaffold's structural stability as S2. Realizing other materials S3 helps strengthen structural solidity not hindering original dynamic functions of S1 onto Scaffold; mixing traditional Materials S3 with S1, effect of F2 can strength F1 then help achieve the desired structural stability.

Table 6.
Three levels of SFA with possible application on 3DPVS.

5.2.3 SFMA into analytical model of system conflicts (AMSC)

From the three-class SFA there are several models of typical conflicts for System Contradictions, which could be used in design methodology. These models are called AMSC. Due to study, 11 AMSC could have practical use for 3DPVS context. In this

No.	AMSC	Description and analysis	Direction on 3DPVS
1	Counterwork	Positive Action or function from S1 to S2 is accompanied by the reverse negative or harmful action or function.	Scaffold holds cells and cells gradually damage scaffold.
2	Mated function A	Positive Function from S1 to S2 is combined with a negative function from S1 to S2, that is, same function can be both positive and harmful at different stages or aspects.	Vibration can accelerate cell growth while damage cells.
3	Mated function B	S1 generate positive function to part 1 of S2, while damage part 2 of S2.	For vibratory cell culture, a definite vibration can help cell proliferation but hinder gene expression.
4	Mated function C	For a S1, S2-A, S2-B system, function from S1 helps S2-A, while hinders S2-B.	Heat from 3DP can melt one material of the material mixture, but damage another.
5	Self-compensating function	While S1 generate useful function to S2, it generates negative aspects on itself.	Some magnetic materials act to magnetic field vibration while their magnetic power is running out. It needs to be charged during the period or after use.
6	Incomplete or Absent function	There are three types. First, S1 generates several useful functions but one or more is absent; second, S1 generates one useful function and strength of which is weak, and third means only product S2 exist while Tool S1 is missing.	Vibration could help proliferation but does little on differentiation; Vibration does not generate sufficient frequency; traditional 3D static scaffold which lacks vibratory stimulation;
7	Insufficient Function	Like the fifth model, useful function from tool S1 to product S2 is not enough.	Cells require a 500HZ frequency of vibration, but mechanical stimulator can generate 300 at maximum.
8	Excessive or Overloaded Function	There are excessive actions generated from S1 to S2 than required.	Mechanical vibration tends to be rough compared with subtle vibration means;
9	Uncontrollable or low-controllability Function	S1 generates useful action to S2, but it is not controllable.	Vibration generated is continuous, while cells need periodic vibration.
10	Incompatible Function	Two tools (S1-A and S1-B) generate useful functions for product (S2), while these two functions contradict or hinder each other.	Definite scaffold's material composition helps cell growth and definite pore structure of scaffold helps cell growth, while the material cannot be fabricated into that pore structure.
11	Mutual-damaging Function	S1 generates harmful action to S2, and S2 have negative effects on S1.	Scaffold hinders the movement of fluid medium for cell culture, while the fluid can corrupt some material of scaffold.

Table 7.
Description of AMSC and regarding examples on 3DPVS scenario.

section, we will illustrate these AMSCs with description, analysis as well connecting with 3DPVS scenario. Using these AMSCs, identifying types of problems and possible solutions tend to be easier and effective. Analyzing AMSC with potential ISS, which will be discussed later, designers could practically find innovation concepts for most design issues (**Table** 7).

6. Innovation standard solutions (ISS) assisting 3DPVS design

TRIZ Innovation standard Solutions (ISS) have been proved as powerful tool for product innovation and novel design [29, 30]. As illustrated in the four-route methodology, ISS could help solving contradiction especially in Route three, where contradiction could not be easily identified as technical contradiction or physical contradiction. Besides this, ISS could be used separately as a philosophy of general Innovation thinking, which could help designers be more practically creative.

6.1 Knowledge and concerns of the dual innovation standard solutions

In recent years, Oxford Creativity, by Karen Gadd and her teams, updated TRIZ Innovation standard solutions into new formats in order to make it relatively easier to implement [29]. Both types of solutions could benefit the design purpose, and in this study, we will firstly learn from the philosophy and then adjust the solutions tailored for 3DPVS design context.

Altshuller and his associates generated the "76 Standard Solutions" in TRIZ between 1975 and 1985. These solutions were grouped into five categories as shown (a) **Table** 8. Despite its high popularity, there are several concerns when trying to utilize them into the 3DPVS innovation context. Firstly, SFA connected with this solution system offers innovative solutions, but some of them tend to be difficult to apply when solving problems. For a given problem, researchers sometimes need to read through every principle to find the suitable one, that is, the traditional 5-class categorization might not be effective in quickly indicating the right direction for problem-solving. On other hand, some solutions could be too general, and others seem to be too specific and focused only in one application. For example, ferromagnetic was the novel product 30 years ago, so TRIZ used it as one of innovation solutions; but taking this to current innovation world is narrowed, that is, "ferromagnetic solution" needs to be put together with antiferromagnetic, diamagnetic, ferromagnetic, and paramagnetic, so to make this solution more generalized and practical. In this regard, for the tailored ISS used for novel design, especially 3DPVS, while the core philosophy remains, some of the solutions might need to be adjusted or rearranged, for more effectiveness and efficiency.

On other hand, studying Standard Solutions Adjusted by Oxford Creativity is also important for the restructuring work into 3DPVS context. Oxford Creativity restructured a new Standard Solution System from the traditional one, aiming to pattern the 76 principles into easily understandable formats so more people can use it quickly. Its classification is shown in (b) **Table** 8. The new standard solutions provide with high understandability, especially for those who are not familiar with TRIZ and those who want instant solutions. However, there are several concerns in this categorization. Before adjusting their solutions to ISS of 3DPVS, understanding this is considered as necessary. That is, first, TRIZ solutions focus on system development, which

a. Innovation Principles by Traditional TRIZ			
Class No.	**Description**	**Notes**	**3DPVS Applicability**
1	**Improving the system with no or little change**	13 Standard Solutions	Yes, high
2	Improving the system by changing the system	23 Standard Solutions	Yes, high
3	System transitions	6 Standard Solutions	Yes, high
4	Detection and measurement	17 Standard Solutions	Yes, low
5	Strategies for simplification and improvement	17 Standard Solutions	Yes, low
vs.			
b. Innovation principles by Oxford Creativity adjusted TRIZ			
Class No.	Description	Notes	3DPVS Applicability
1	Harm	24 Solutions with 4 Sub-class	Yes, high
2	Insufficiency	35 Solutions with 3 Sub-class	Yes, high
3	Measurement	17 Solutions with 3 Sub-class	Yes, low

Table 8.
(a) Innovation standard solutions by traditional TRIZ vs. (b) Adjusted innovation principles by Oxford creativity.

contains sub-system components, system, and super-system. Oxford Creativity does not prioritize in system. And in this connection, some definition of solutions might be vague. For example, formula of solutions as, "1.1 Add something to/inside the subject or object," and "1.3 Use the environment," which seems too general and not practically helpful. Also, some definition of the class seems debatable. For example, to "Eliminate, trim out the harm" occupies six solutions, and to "stop, block the harm" occupies another 11. The two classifications however describe the same Harm-elimination process, just in different level or degree. Eliminating harm means fully stopping and partially stopping means eliminating. Further concern is regarding Field and Action. In TRIZ, action is a by-product resulted from the reaction of F, S1 and S2, basically the consequence of elements or process generated by elements. Field and Action are not the "phenomenon" in one level. Field can contain an action, but not action itself. Additionally, the congruity could also be a concern. For example, the first two solution classes of Oxford Creativity, namely Harm and Insufficiency, are based on functions between field F and Substances S1, S2, while the third one, namely measurement, is based on requirements by researchers on the product level or on the function level of product. For another example, for the Measurement, there includes two Functions as well, harmful means of measurement or insufficient means measurement, then solutions given in Measurement category seem as repetitive with solutions given in first two categories. In practical use, this might cause repetitiveness in solution-searching [29]. Some standard solutions are partly repetitive with 40 principles. Therefore, rapid solution-searching while maintaining the general congruity is needed. These concerns might need to be addressed, to a certain extent, in the ISS established for 3DPVS design.

6.2 Restructuring ISS for tailored 3DPVS design context

To assist with the four-route methodology established by this research for the novel development of 3DPVS, we attempt to adapt and restructure the Innovation Standard Solutions (ISS) based on Altshuller's 76 Standard Solutions and the simplified version of Oxford Creativity.

As early introduced, the essence of 3DPVS development is a cross-domain scientific design covering realms of scaffold engineering, vibration science, mechanical science, 3DP, and so forth. Therefore, standard solutions or principles need to be tailored, or at least compatible for these realms, to ensure with design proficiency and convenience. The proposed ISS contains most of the original standard solutions in both traditional TRIZ and Oxford Creativity, with new extensions and modifications; that is, part of the TRIZ and Oxford solutions have been simplified to make it more scientifically specific and accurate, part have been modified to be general and applicable for wider context, and several derived new solutions have been added covering novel cross-domain innovation aspects. In brief, ISS for innovation of 3DPVS, as well other similar bio-product design, is proposed to make best of both worlds of traditional TRIZ and Oxford Creativity while addressing some limitations they potentially have on 3DPVS context. Traits of ISS include but not limited to following aspects: first it can be directly anchored with the 40 Innovation Principles for solving 3DPVS System Contradiction; second, it potentially provides with more effectiveness, clarity, and integrity when cooperating ISS with SFMA methods; and third, new formula of most solutions has been uniformed: "what action"+ "how to act"+ "where/when is the context"; this makes each solution be understandable, specific, and accurate. In this connection, four classifications with be established, with the philosophy and focus as follows.

Contradiction-solving-based. In this category, five Solutions revolve around solving concerns related with consciousness and contradiction-solving. Consciousness from one engineering realm toward others, especially cross-domain ones, seems always challenging and understanding, addressing this is thus necessary for every designer. Since contradiction is the core of problem-solving in TRIZ, better understanding and analysis on contradiction and make it applicable, become another vital issue. It is also worth noting that before analyzing the solutions of usefulness, or harm, which will be discussed later, the system model itself needs to be established properly in accurate way; otherwise the following analysis to tackle "harm" or strengthen "usefulness" will be ineffective or even faulty. The process of establishing a proper SFMA is predominant, and it can be considered as "Neutral" in terms of the desired or undesired functioning of reality system, which is represented by "usefulness" and "harm" respectively.

Usefulness-based standard solutions. In this category, eight Solutions concern with Useful Action, function, or Activity that is needed by system while seems lacking, incomplete, or inadequate. First, four solutions will be about "Creation of SFM from incomplete model"; second, two solutions are regarding the "Delete excessive elements of SFM, and making it simplified"; third, one solution focuses on "Transform traditional SFM into NSFM which contains the analysis of Three Forces."

Harm-based standard solutions. As the opposite of "usefulness," "harm" needs to be mitigated, decreased during design process. This category contains 26 elementary solutions with two basic strategies in problem-solving. First, 18 solutions would deal with "Eliminate the Harm or Decrease its Degree" and secondly, eight solutions focus on "Transform harmful functions, activities or objects into positive."

Similar as Usefulness-based solutions, "harm"-based solutions could potentially be applied not only on technical engineering aspects for 3DPVS but also useful to solve other harmful actions or functions as it represents the very H force in Law of Three. These solutions could potentially be the universally standard solutions to the engineering, scientific, and philosophic world, which deals with "Harm" and "Usefulness."

Insufficiency-based elements. Two basic strategies composing 37 solutions are included in this category, which aims to help improve, change, or enhance definite functions by changing the Substances (S1 or S2), or the action/ field (F), which acts between them. Nineteen solutions deal with the "Change, evolve the system" components or add new elements to them. Eighteen solutions focus on "Enhance, strength or improve the Action focusing on Insufficient or Missing Field." In this category, to find the right F when there is an object (S2), which needs to provide some extra functions or change, a subject or tool (S1) with a field (F) to deliver/complete such function, becomes the key. This group of standard solutions also helps find a solution, particularly when cooperating with Scientific Effects Analysis.

To conclude, ISS established inside TRIZ-based methodology could increase the speed of generating innovative ideas and creating novel but practical concepts, which is especially necessary in the 3DPVS conceptual developing scenario. ISS could not only be used for Route 3 and 4 where no easy identification on System Contradiction or Physical Contradiction of definite design question can be obtained, but also as general innovation standards for much wider multi-cross domain designs and innovations.

7. Concluding remarks

In this study, TRIZ methods have been reviewed systematically and exploring them for the concept design of 3DPVS has been illustrated and discussed. Following this study, a proper methodology derived from TRIZ could be created. Considering the possible limitations of traditional design approaches on scaffold engineering and the limitations on TRIZ, we tailored and transferred traditional TRIZ-based principles into the new context that can be directly applicable by concept generation of 3DPVS. For several key information, for example, the generated concept criteria, rank scores, and selected optimal concept, readers can refer to author's another paper [31] for more details.

Further connected, this study aims to provide a useful, effective, and accurate cross-domain aspects of TRIZ for the innovation and design for novel scaffold. In this connection, 39 Parameters TRIZ Matrix and 40 Innovative Principle for system contradiction could be attempted with core elements transferred into scaffold contents especially tailoring for 3DPVS conceptual design. This means that, original explanation of TRIZ principles and parameters, their possible and extended indication, as well as the application on 3DPVS could be analyzed comprehensively. SFM and SFMA have been studied and proposed with their potential new functions, with the Three Forces embedded. Next to this, from traditional TRIZ-based ISS and the newly brought ones, a same-essence but restructured format can be generated, conveniently dealing with the comprehensive aspects appearing on innovating 3DPVS. The ISS for 3DPVS coming from the Innovation principles, making SFM process more easily addressable.

This study has established a ready-to-use knowledge base, which could connect TRIZ innovators and those who focus on novel 3DPVS designs but do not know how to use TRIZ. Utilizing TRIZ-based method, it is expected that the conceptual design, especially the concept generation, will take place more efficiency. Thus the study can fill a gap in the scaffold design and TRIZ literature, which are preliminary to preparing future works of establishing formal TRIZ-based methodology for scaffold innovation.

In terms of the future work following this study, it may include a validating part on the Expert Survey, which can be conducted on experienced experts in relevant but cross-domain academic realms. Analyzing experts' feedback, we might draw a preliminary conclusion about the basic relevance, usefulness, and practicality of TRIZ-based methods on 3DPVS conceptual design.

Compliance with ethical standards

Conflict of interest

The authors declare no conflict of interest.

Ethical approval

This study does not contain any studies with human or animal subjects performed by any of the authors.

Author details

Haobo Yuan
School of Engineering, University of South Australia, Mawson Lakes, Adelaide, Australia

*Address all correspondence to: yuahy006@mymail.unisa.edu.au

References

[1] Yuan H, Xing K, Hsu H-Y. Trinity of three-dimensional (3d) scaffold, vibration, and 3D printing on cell culture application: A systematic review and indicating future direction. Bioengineering. 2018;**5**:57

[2] Yuan H. Introducing the language of "relativity" for new scaffold categorization. Bioengineering. 2019;**6**:1. DOI: 10.3390/bioengineering6010020

[3] Yuan H, Xing K, Hsu H-Y. Concept justification of future 3DPVS and novel approach towards its conceptual development. Designs. 2018;**2**:23

[4] Finger S, Dixon JR. A review of research in mechanical engineering design. Part II: Representations, analysis, and design for the life cycle. Research in Engineering Design. 1989;**1**:121-137

[5] Blogg, K. The Selection and Application of Design Methodologies for the Design of Bone Tissue Scaffolds. PhD thesis, Cranfield University; 2013

[6] Papenburg B. Design strategies for tissue engineering scaffolds [PhD thesis]. The Netherlands, Bernke Papenburg, Enschede, The Netherlands: University of Twente; 2009

[7] Fiorineschi L, Frillici FS, Rotini F, Tomassini M. Exploiting TRIZ Tools for enhancing systematic conceptual design activities. Journal of Engineering Design. 2018;**29**:259-290. DOI: 10.1080/09544828.2018.1473558

[8] Shah JJ, Kulkarni SV, Vargas-Hernandez N. Evaluation of idea generation methods for conceptual design: Effectiveness metrics and design of experiments. Journal of Mechanical Design. 2000;**122**:377-384

[9] Li X, Li L, Chen Z. Toward extenics-based innovation model on intelligent knowledge management. Annals of Data Science. 2014;**1**:127-148. DOI: 10.1007/s40745-014-0009-5

[10] Bogatyrev N, Bogatyreva O. TRIZ-based algorithm for biomimetic design. Procedia Engineering. 2015;**131**:377-387. DOI: 10.1016/j.proeng.2015.12.417

[11] Kamps T, Christopher M, Stacheder L, Seidel C, Reinhart G, Lindemann U. TRIZ-based biomimetic part-design for Laser Additive Manufacturing. Lasers in Manufacturing Conference 2015. Munich; 2015

[12] Sarkar P, Chakrabarti A. Assessing design creativity. Design Studies. 2011;**32**:348-383

[13] Altshuller GS. Creativity as an Exact Science : The Theory of the Solution of Inventive Problems. New York: Gordon and Breach Science Publishers; 1984

[14] Bukhman I. TRIZ Technology for Innovation. Cubic Creativity Company: Taipei; 2012

[15] Chechurin L. Research and Practice on the Theory of Inventive Problem Solving (TRIZ) : Linking Creativity, Engineering and Innovation. Cham: Springer; 2016

[16] Altshuller GS. In: Shulyak L, Altov H, Dronova N, Urmanchev U, editors. And Suddenly the Inventor Appeared : TRIZ, The Theory of Inventive Problem Solving. New ed. Massachusetts, US: Technical Innovation Center; 2004

[17] Cascini G, Russo D. Computer-aided analysis of patents and search for TRIZ contradictions. International Journal

of Product Development. 2007;**4**:52-67. DOI: 10.1504/ijpd.2007.011533

[18] Blogg, K. The Selection and Application of Design Methodologies for the Design of Bone Tissue Scaffolds; 2013

[19] Ilevbare IM, Probert D, Phaal R. a review of triz and its benefits and challenges in practice. Technovation. 2013;**33**:30-37. DOI: 10.1016/j.technovation.2012.11.003

[20] Jia-Yen H, Shih-Tian S. Development of an integrated bionic design system. Journal of Engineering, Design and Technology. 2016;**14**:310-327. DOI: 10.1108/JEDT-08-2014-0057

[21] Cameron G. TRIZICS : Teach Yourself TRIZ, How to Invent, Innovate and Solve "Impossible" Technical Problems Systematically. CreateSpace: United States; 2010

[22] Sekar V, Jayakrishna K. Development of integrated ECQFD, LCA and sustainable analysis model. Journal of Engineering, Design and Technology. 2014;**12**:102-127. DOI: 10.1108/JEDT-07-2011-0052

[23] Fouchal F, Hassan TM, Loveday DL. Design methodology for integrating multipath systems (building services). Journal of Engineering, Design and Technology. 2014;**12**:491-506. DOI: 10.1108/JEDT-09-2011-0057

[24] Dorst K, Cross N. Creativity in the design process: Co-evolution of problem–solution. Design Studies. 2001;**22**:425-437. DOI: 10.1016/S0142-694X(01)00009-6

[25] Chakrabarti A, Bligh TP. A scheme for functional reasoning in conceptual design. Design Studies. 2001;**22**:493-517. DOI: 10.1016/S0142-694X(01)00008-4

[26] De Carvalho M, Back N. Cross-Fertilization Between TRIZ and the Systematic Approach to Product Planning and Conceptual Design. Worcester/MA: Altshuller Institute; 2000

[27] Fey V, Rivin E. Innovation on Demand: New Product Development Using TRIZ. Cambridge, UK: Cambridge University Press; 2005

[28] Altshuller GS. In: Shulyak L, Rodman S, Clarke DW, editors. 40 Principles Extended Edition: TRIZ Keys to Technical Innovation. Massachusetts, US: Technical Innovation Center; 2005

[29] Goddard C, Gadd K. TRIZ for Engineers Enabling Inventive Problem Solving. Chichester, West Sussex, U.K, Hoboken, N.J: Wiley; 2011

[30] Savransky, S.D. Engineering of Creativity: Introduction to TRIZ Methodology of Inventive Problem Solving. CRC Press; 2000. ISBN 9780849322556

[31] Yuan H, Xing K, Hsu H-Y. Generate basic conceptual solutions for 3DPVS via utilizing TRIZ. Bio-Design and Manufacturing. 2019;**2**:76-95. DOI: 10.1007/s42242-019-00039-6